The Energy-efficient Church

The Energy-efficient Church

How to Save Energy (and money) in Your Church

Total Environmental Action, Inc.
Douglas R. Hoffman, Editor
Foreword by Gordon Bucy, A.I.A.

The Pilgrim Press • New York • Philadelphia

Library of Congress Cataloging in Publication Data

Total Environmental Action, Inc.
The energy-efficient church.
Bibliography: p. 85
1. Church maintenance and repair. 2. Churches—
Energy conservation. I. Hoffman, Douglas R., 1949-
II. Title.
BV652.7.T6 1979 254'.7 79-10432
ISBN 0-8298-0362-9

The Pilgrim Press, 287 Park Avenue South, New York, New York 10010

Contents

Foreword

By Gordon Bucy, A.I.A.

Every organizational expression of the church must place the subject of energy on its agenda for the remaining part of this century. Each of us must assess the opportunities to turn that agenda into constructive action.

We need voices that speak the truth about our past and the opportunities in our future. When a total perspective of human life is viewed within measurements of time and consumption, the last century reveals careless, selfish, and arrogant attitudes toward the stewardship of Earth's resources. High technology has promised and delivered so much that prophetic murmurs about corruption in Earth's management get muffled by the cries for unbridled consumption.

Our contemporary society has placed greater values on consumption rather than on conservation. We should observe by now that a throwaway society winds up as an ever-expanding garbage heap. No matter what we do to cover it over with trees and grass, our conscience tells us that using less and recycling nonrenewable resources is a better way to show responsibility to future generations.

The number of persons dedicated to sensible use of natural resources is growing. Some are capable of giving direction, as illustrated by the authors of this book, and others are capable of implementing many solutions that counter the excessive consumption habits of our society. Whether your motive is survival for yourself, your favorite institution, the whole world, or a conscientious effort to acknowledge God's gifts in a proper way, now is the time to take those small steps that represent struggle and hope for the future.

Gordon Bucy is the Executive Vice President of the American Baptist Extension Corporation in Valley Forge, Pennsylvania.

Acknowledgments

This book was prepared by Total Environmental Action, Inc., Harrisville, New Hampshire for the Office of Architecture of the National Division, the Board of Global Ministries, United Methodist Church. Those particularly responsible at Total Environmental Action include Barbara Putnam, author and illustrator; Mona Anderson, editor; and Duncan Bremer, author and project manager. Many other staff members also contributed. The introduction was written by Gordon Bucy, of the American Baptist Extension Corporation. The project was sponsored by Douglas R. Hoffman, of the Office of Architecture (United Methodist Church), with the support and encouragement of the Church Development Task Force of the Joint Strategy for Action Committee (JSAC), an ecumenical task force focusing on new church development and the redevelopment of existing congregations. We would also like to acknowledge the assistance given by the Energy Education Project of the Division of Church and Society, National Council of Churches.

The Energy-efficient Church

Energy Conservation Principles:

Energy Conservation Principles

The Energy Crisis

The use of energy in the United States has been growing yearly at a rate of 4 percent or more. Ninety-three percent of this energy is supplied by fossil fuels: coal, oil, and natural gas. Since 1968, crude oil prices have risen from less than $3 a barrel to more than $10 a barrel, and coal prices have risen from $5 to $20 a ton.* Fuel is scarce, and the costs of supplying this energy, both economic and environmental, are growing.

Our country uses more energy than it produces; our oil imports have more than doubled since 1968, and now 42 percent of our oil must be purchased abroad. This dependence on foreign imports damages our balance of payments and causes inflation.

The economic pressure to supply more energy often obstructs efforts to protect the environment. The environmental costs of strip mining, offshore drilling, oil spills, air pollution, and nuclear waste disposal are not easy to measure but are certainly part of the price we pay for our energy.

---

*1976 Figures, Federal Energy Administration, *Energy in Focus,* FEA/A-77/144, May 1977.

Energy Conservation

We are finally realizing that fossil fuel reserves are limited, not infinite; we could run out of them before the end of the century. At this point, however, *when* they run out is still our choice. If we reduce our use of these resources, they can last through several more generations, giving us time to develop alternative, renewable energy sources. On the other hand, if we continue consuming fossil fuels at the present rate, supplies could be exhausted in our own lifetimes. Energy conservation will do more than reduce individual fuel bills. If practiced on a national scale, it will also reduce economic pressures that encourage environmental pollution and promote inflation. More importantly, energy conservation will help to preserve our limited resources for future generations.

Churches and Energy Conservation

Churches present several unique opportunities for energy conservation. First, church buildings are subject to intermittent use. The largest rooms, for example, are probably used infrequently, and yet these rooms are often heated constantly. Second, churches can mobilize volunteer labor to implement energy-saving measures. Third, churches have an extended time perspective and frequently think in terms of investments in the future. Fourth, churches can usually borrow money for energy conservation measures at favorable interest rates. Finally, churches recognize the responsibility for leadership in matters of community importance.

The Energy-efficient Church

The Energy-efficient Church seeks to point out these unique opportunities and to raise the energy awareness of church members. Primarily, this book is a guide for reducing fuel use in church buildings. Its purpose is to encourage church members and administrators to take action. It describes energy conservation possibilities that are particular to

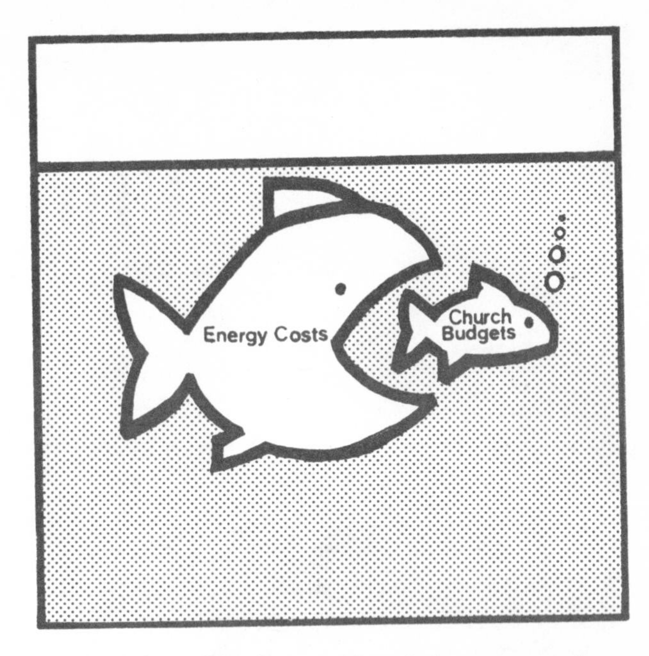

large, intermittently used buildings as well as those that apply to all buildings. In addition, guidance is offered in choosing the measures that will be most appropriate for your particular church plant.

Choosing the Right Energy Conservation Measures:

Choosing the Right Energy Conservation Measures for Your Building

Making Decisions

The circumstances in which each of the following energy conservation measures is likely to be most appropriate are discussed below. Many factors affect the use of fuel: type of building construction, size and use patterns of the building, type of heating and cooling system, type of fuel used, and local climate. Those factors that have the greatest effect on the possible fuel savings are noted under each measure. Oftentimes this may be enough information on which to base a decision. In other cases, there will be a number of measures to choose from, and you will need to decide which to pursue first. The concept of cost-effectiveness will be useful in making these decisions.

Cost-effectiveness

If two energy conservation measures have the same initial cost but one will save 20 percent of the energy used and will last twenty years and the other will save 10 percent and last ten years, clearly the one that lasts longer and saves more is preferable. The measure that saves the most per dollar spent, over its lifetime, is the most *cost-effective*. Sometimes the cost-effectiveness of an energy conservation measure will be clear; measures that save a great deal and cost very little are obviously cost-effective. In other cases it will not be as evident; for example, if it is necessary to borrow money, how does one account for the additional cost of interest? If there are annual maintenance expenses, such as the cost of putting up and taking down storm windows, how does this alter their cost-effectiveness? Will a large capital expense be justified in the long run? In each of these instances, it may be worthwhile to calculate the measure's cost-effectiveness, using the life cycle costing method shown in Appendix A. The result of the life cycle costing calculation is a benefit-cost ratio. The benefit-cost ratio represents the present value of savings (in dollars) for each dollar spent over the lifetime of the energy conservation measure. If benefit-cost ratios are calculated for a number of feasible measures, it is then possible to compare them and choose the most cost-effective, the one with the highest benefit-cost ratio.

Since the benefit-cost ratio represents dollars saved over dollars spent, the cost-effectiveness of long-lasting measures with a high initial expense can be compared with short-lived measures that cost less. If their life cycle is long enough, even very expensive measures may prove to be as cost-effective as cheaper ones that don't last as long.

In order to calculate the cost-effectiveness of an energy conservation measure through the life cycle costing method, you will need to know the measure's initial cost, maintenance cost, life cycle, and estimated annual energy savings. The calculation method given in Appendix A requires an estimate of the cost of borrowing money (interest) and of the inflation rate of fuel. Measures that have annual maintenance expenses also require an estimate of the general inflation rate.

Ranking

Because the amount of fuel saved by each measure will vary from building to building, this book cannot provide exact estimates. Instead, it gives each measure a rank, based on the range into which benefit-cost ratios generally fall. To help you calculate an exact benefit-cost ratio, a list of the factors affecting energy savings and of the sources for securing further information are given.

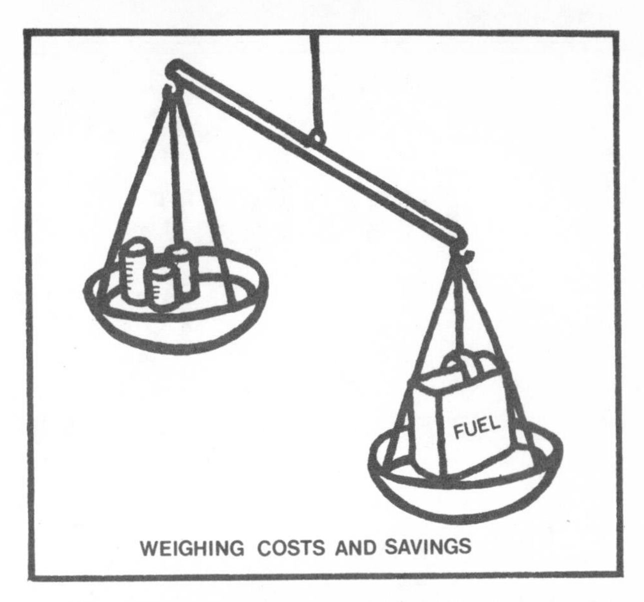

WEIGHING COSTS AND SAVINGS

Frequently, however, an initial decision can be reached on the basis of the rank, without calculating the benefit-cost ratio. This ranking scheme will help reduce the number of life cycle costing calculations required and will therefore simplify the decision-making process.

The ranking scheme is as follows:

HIGH: Do it without weighing costs and benefits. Costs are generally very low, benefits very high.

HIGH MIDDLE: Do it if the costs are not substantial. Benefits are generally very high. If costs are substantial, weigh costs and benefits carefully.

LOW MIDDLE: Do it after weighing costs and benefits carefully. Costs may be substantial, but so may benefits.

LOW: Do it after everything else is done. Costs are usually high in proportion to benefits, but special circumstances, such as extreme climate or high fuel costs, may justify the expense.

It may be necessary to calculate the benefit-cost ratio only when a choice must be made between two measures with the same rank.

Energy Conservation Plan

While costs and savings are the basis for evaluating the relative merits of energy conservation measures, other factors are significant when actually putting energy conservation into practice. The measures suggested in this book involve several types of changes: changes in the building's mechanical and control equipment, changes in the way the building is used, and changes to the building shell itself. Some cost money: others simply require that building users be more energy conscious. Some can be done by church members; others may indicate a need for hiring experts to determine cost-effectiveness. Some involve a one-time effort; others demand constant attention. Some necessitate a finance committee decision; others mean a change in the routine of the sexton or church secretary.

To assure the success of such a multifaceted program it will be necessary to draw up an energy conservation plan, which should include a list of the energy conservation measures appropriate to your church in order of their rank and the type of action required to effect them. This plan can be used to inform church members and as a reference whenever new resources for energy conservation become available. It will clarify which actions should be taken immediately. Measures that are less cost-effective or are more expensive can be deferred for a later time. The plan also serves as a guide for future decision-making. This manual will enable you to formulate a list of energy conservation possibilities for your building, along with a preliminary rank and estimate of resources required to implement or further evaluate each one. A sample format for such a plan is shown.

ENERGY CONSERVATION PLAN

MEASURE	RANK	ACTION REQUIRED
set back thermostats	high	inform members, sexton
tune up furnace	high	—get cost estimates —get authorization from budget committee —inform sexton
insulate attic	high middle	—get cost estimates —estimate fuel savings —recruit volunteers for fundraising, labor

COST-ESTIMATING

MEASURE	ESTIMATED SAVINGS	COST	RANK
1.			
2.			
3.			
4.			
5.			
6.			
7.			
8.			
9.			
10.			
TOTALS:			

Perhaps the most essential part of an energy conservation program is educating those who use the building. If the users understand the reasons for changes, they will be more willing to cooperate. User awareness can effect real savings. One study has shown that 10 percent of the savings in energy conservation programs is a result of the users' new awareness that energy should be saved.

Energy conservation is an ongoing process, not a single act. Consideration should be given to having someone act as an "energy watchdog" to monitor fuel and electrical use and to attend to the various energy conservation changes.

As fuel becomes scarcer and energy prices rise, the need for conservation will increase. Measures that are not currently cost-effective will become cost-effective as fuel costs rise. However, saving money is not the only reason for saving energy. Your church may feel it should err on the side of saving "too much" energy in order to demonstrate leadership in the community. Energy conservation is an act of good stewardship of the riches of creation. Working together to save energy can be an act of fellowship.

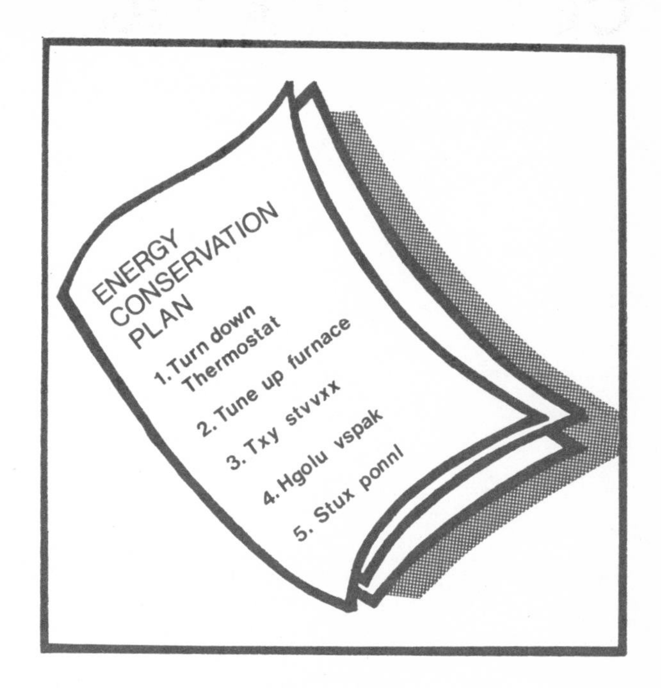

Energy Conservation Measures:

Energy Conservation Measures

The energy conservation measures discussed in the following sections range in order from the highest benefit-cost ratios to the lowest: first, *Reducing Inefficiency;* second, *Reducing Demand;* and third, *Reducing Building Load.* Every conservation measure saves energy. The greatest portion of wasted energy, however, occurs in an inefficient heating or cooling plant and through heating or cooling unused portions of a building. In addition, the measures detailed in the first two sections cost relatively little. The combination of low cost and high benefit produces a high benefit-cost ratio, so concentrate on these two areas first. The measures in the third section, *Reducing Building Load,* are generally less cost-effective but still may be worthwhile. When a major expenditure is involved, estimates of energy savings and costs should be prepared. These estimates can be used to make a life cycle cost analysis to determine which measure should be implemented first.

There is a basic relationship between climate and energy use: the more severe the climate, the more energy is used. A similar relationship exists between energy use and the energy saved by conservation measures: the more energy presently used, the greater the potential savings. Therefore, the same conservation measure practiced in two different churches will save more energy in the larger church. Similarly, the same conservation measure applied in two different churches will save more energy in the church situated in the more severe climate. These churches, then, should expect to spend more money and effort on energy conservation.

Building codes and conditions vary from place to place and even from building to building. Although it is desirable to use volunteer labor where possible, installation should conform to good practice and applicable codes. This means there should be adequate supervision and a thorough review with the building inspector. Improperly installed insulation, for example, could become a fire hazard. Or thermal curtains or shutters might obstruct exits or be made of dangerous materials. Even though information on the energy conservation measures themselves is provided, the specific application of the measure to a particular church building cannot be anticipated.

An energy conservation analysis of three Massachusetts churches, conducted by Total Environmental Action, Inc., provides cost and savings information throughout the book. This investigation showed that these churches could often realize savings of up to 50 percent of their energy use (often thousands of dollars per year) for only hundreds of dollars in investment. The figures used for these particular churches are from 1977. Costs and benefits will differ for an individual church, and they are certainly affected by inflation. For more information on the study, consult the Energy Conservation section of the Bibliography.

Generally, the study indicated that each of the three churches should have been able to cut its energy consumption 9% to 36% by reducing inefficiency. Initial costs would be $450 to $775, with annual savings of $160 to $1,600 per year. By reducing demand, these churches should have been able to cut consumption an additional 16% to 37%, with first costs of $150 to $3,000 and annual savings of $300 to $7,200. Finally, by reducing building load, they would save an additional 20% to 40% at costs of $1,800 to $6,700 and annual savings of $340 to $2,700.

Remember that these figures are for specific churches in a specific climate. Use them only as a guide. The information in the following sections will help determine the right measures for *your* church building.

Don't Send It up the Chimney: Reducing Inefficiency

A building's heating plant converts fuel to useful heat. The process is twofold: the heating plant burns fuel, and the resulting heat is distributed throughout the building. During either step, heat energy can be wasted through inefficiency. Because the heating plant handles *all* the energy used to heat the building, even a small improvement in its efficiency will pay large dividends. Fortunately, many improvements are relatively low-cost. The heating plant, therefore, is the first place to examine in an energy conservation program.

Efficiency is a measure of the useful energy produced by the system. In a heating plant, efficiency equals the percentage of heat energy released by the fuel that remains in the system and is circulated to the building. The rest is wasted heat, going up the chimney as hot flue gases. A certain amount of waste is unavoidable, but proper maintenance of the heating plant will minimize it.

The efficiency of the heating plant can be controlled as long as fuel is burned in the building. If the heat comes from electrical resistance heating, the efficiency with which it is generated and transported is in the hands of the electric utility, and this section will not apply.

Heat produced in a heating plant is the result of the combustion of fuel. In order to burn, the fuel needs to mix with the proper amount of air. Too little air will cause incomplete combustion, and the unburned fuel will appear as smoke and soot. Too much air will cause the heat to shoot up the chimney before it has a chance to heat the building.

Once heat is produced, it is transferred through a heat exchanger to a heat transport

medium—air, water, or steam. In order for this heat transfer to be efficient, the surfaces of the heat exchanger must be clean.

The heat transport medium is then circulated to the rooms in the building. Air systems use a fan or blower to circulate the air through ducts to registers in each room. Pumps circulate hot water through pipes to radiators, and steam moves by its own pressure to the radiators.

If the heating plant is not burning fuel efficiently, the heat transport medium will not have as much heat to circulate and will therefore have to circulate longer. The burner, too, will have to fire for longer, more frequent periods. Improving the efficiency of the heating plant will not reduce the building's need for *heat,* but it will reduce the building's need for *fuel;* it will take less fuel to make the required heat. For this reason, burner efficiency is the first measure in this section.

The burner does not fire all the time; it comes on when the temperature or pressure of the heat transport medium has dropped to a certain level. At the beginning of each cycle, when the burner is still warming up, it operates at less than peak efficiency. After the burner has shut off, the draft that supplies the air needed for combustion continues to pull air through the heat exchanger and out of the building. Consequently, the unavoidable on-and-off cycling of the burner reduces the overall operating efficiency. The second and third measures in this section reduce this waste. First, excessively frequent cycling can be rectified by altering the burner. Second, the draft that robs the heat exchanger of heat during the burner's off cycle can be controlled by a motorized damper.

In addition to burner efficiency and cycling, the way the heat travels to the rooms requires attention. Maintenance of the distribution system is essential to the overall efficiency of the building's heating plant. These four measures will help keep the heat from going up the chimney.

Burner Efficiency

The first thing to check in the heating plant is the combustion efficiency of the burner. The standard combustion efficiency tests are CO_2 level, smoke, draft, and stack temperature. CO_2 (carbon dioxide) is a product of combustion, and its level indicates the amount of combustion taking place. Smoke tests show whether or not there is incomplete combustion and if heat exchange surfaces require cleaning. Draft tests determine if there is the proper level of air flow through the combustion chamber. Stack temperatures reveal whether or not too much heat is going up the chimney. A special instrument is needed for each of these tests. In addition, using the results to calculate efficiency and deciding what adjustments have to be made requires training and experience. A heating contractor can perform these efficiency checks and tune-ups, which are routine and relatively inexpensive. They should be done at least once a year; even small inefficiencies waste fuel.

What Are the Savings?

An efficiency check and tune-up will save a higher amount in buildings where such maintenance has previously been neglected. In these cases, there is the most room for improvement.

Since large buildings in cold climates use the most fuel, even a small improvement in efficiency will save a large amount of heat energy. The estimated savings for a tune-up on a large church in the Massachusetts study were 2 percent, resulting in a savings of $240 per year. On a smaller church, replacement of the burner was estimated to cut fuel consumption 27 percent and to save $370 per year.

What Are the Costs?

Heating contractors' fees for testing efficiency, a fairly routine procedure, are about the same for most heating plants, large or small. However, the cost may be higher if the heating plant has more than one burner. The fee for tuning up and cleaning the burner varies, depending on how much maintenance is needed. The Massachusetts study estimated $75 for the testing and tune-up on a

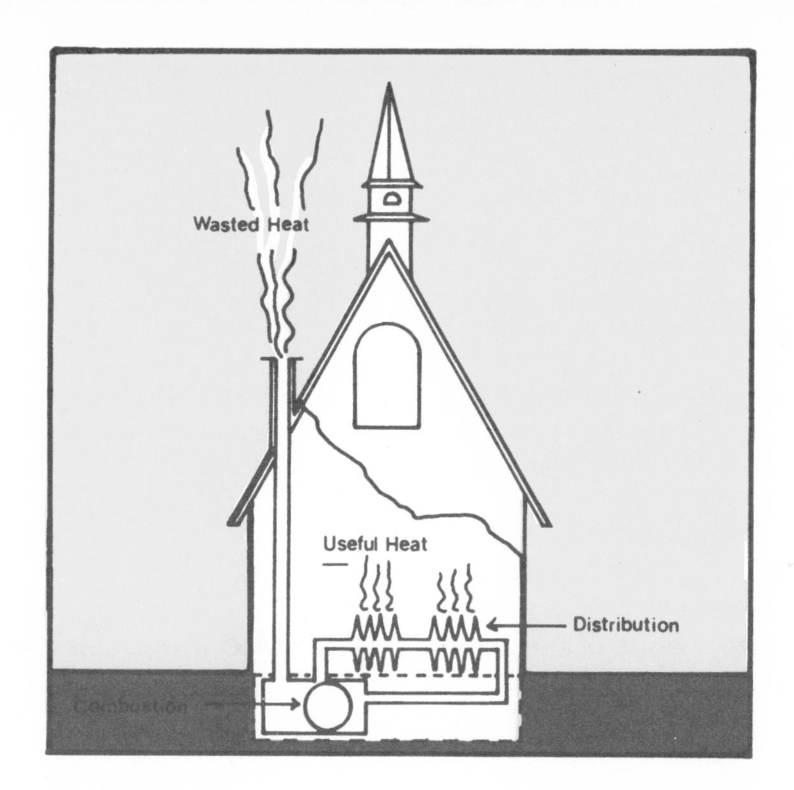

large church. On smaller churches the charge may be as low as $15 to $25.

Can It Be Done by Church Volunteers?

Unless there is a heating contractor in the congregation who offers, this job is not appropriate for volunteer labor.

For More Information

A fuel supplier may do efficiency testing but may not be equipped to completely service the burner. When requesting service, be sure that both an efficiency test and a tune-up are included. A second efficiency check after the burner has been serviced will indicate the degree of improvement. An estimate of fuel savings can be made, based on this change in efficiency.

Rank

HIGH: Do it without weighing costs and benefits.

Frequent Cycling

In most buildings, the heating plant is designed to be large enough to heat the entire building in the coldest weather. In church buildings, all spaces are rarely used at the same time. If unused areas are kept at a lower temperature, the heating system will not have to perform at full capacity to heat the building. This means that the burner will be operating in short bursts, spending much of the time warming up and less time at optimum operating temperature. Also if a building is insulated and winterized, it will no longer need the full capacity of the existing heating plant.

If a building's heating plant is oversized and is cycling on and off too frequently as a result, its operating efficiency may be increased by lessening the amount of fuel introduced into the firebox. This could mean reducing the nozzle size, the size of the burner motor, or both.

Reducing the building's demand and load, discussed on pages 26-33 and 33-51, will decrease the load on the burner. If either of these types of energy conservation measures is to be implemented, a reevaluation of the burner capacity should be made after this implementation.

Like heating systems, air conditioning systems are also frequently oversized. If properly maintained and controlled, and if the building cooling load is lowered, smaller air conditioning systems can provide more efficient cooling. Room air conditioners can be replaced gradually with smaller, more energy-efficient units that will run more continuously and cycle less frequently. Central air conditioning systems should also be investigated. A mechanical engineer or a heating contractor can help with such measures as reducing air flow rates, raising evaporator temperatures and suction pressure, or lowering the condensing temperature.

What Are the Savings?

The greater the oversizing and the annual fuel use of the building, the greater the savings by reducing the capacity of the system. If a system is 50 percent oversized, its overall seasonal efficiency can drop as much as 30 percent.

What Are the Costs?

Costs include both the preliminary expense of hiring a qualified mechanical engineer to evaluate the system and the cost of making alterations to the burner. An engineer should be consulted to determine whether or not the building's actual heat requirements are significantly lower than the rated capacity of the system. The consultant will then advise how the capacity of the system can be reduced and still provide adequate heat. Alteration costs will vary, depending on the system.

Can It Be Done by Church Volunteers?

This is not a job for volunteers, unless one happens to be a qualified engineer.

For More Information

A mechanical engineer or qualified energy conservation consultant can discuss the merits of this option and should be able to give estimates of fuel savings and costs. The assumptions and calculations on which the original system design was based will be helpful in reevaluating it.

For small buildings, consult a heating contractor. For large buildings, *How to Save Energy and Cut Costs in Existing Industrial and Commercial Buildings* may be useful (see Bibliography).

Rank

Consulting an expert ranks HIGH. It definitely should be done. Use her or his cost and savings estimates to rank the alteration work.

Automatic Stack Dampers

The fuel burner does not operate continuously. When the heat transfer fluid is warm enough, the burner shuts off. Remember, however, that even when the burner is not firing, the draft, which was needed to supply air for combustion, continues to pull warm air up the chimney and out of the building. Automatic stack dampers are designed to reduce this waste by automatically closing the stack when the burner stops. The heat remains in the heating plant and is not lost up the chimney. Automatic stack dampers are fairly new devices and can be a fire hazard if improperly designed or installed. Look for those that have been tested and approved by Underwriters' Laboratory, and have them put in place by an experienced installer.

What Are the Savings?

Manufacturers' estimates are in the 10% to 25% range of total fuel use. Because they save heat when the burner is off, automatic stack dampers will save the highest *percentage* of fuel in heating plants that cycle on and off frequently. This type of cycling most often occurs in air systems. They will save the most fuel in buildings that use the most fuel—large, frequently used buildings in cold climates. Savings in the Massachusetts churches were estimated at 9% to 11% ($90 to $1,627) per year.

What Are the Costs?

Costs include both the damper and installation. The cost of the damper will

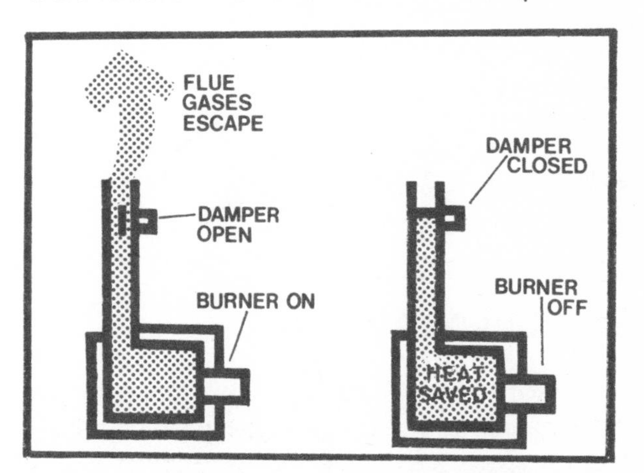

depend on the flue size. The installation cost will depend on the accessibility of the flue and burner controls. Systems with more than one burner will need more than one damper; this will increase the expense. Costs of $200 to $700 were estimated for the Massachusetts churches.

Can It Be Done by Church Volunteers?

Installation of a stack damper is not appropriate for volunteer labor.

For More Information

A reliable heating contractor experienced in installing stack dampers should be able to help choose and estimate installation costs. The contractor may also be able to estimate savings. A manufacturer's engineer can estimate fuel savings if given information on the type of heating system and the building's annual fuel use.

Rank

HIGH MIDDLE: Do it if expense is not substantial. If costs are considerable, compare costs and benefits carefully.

Distribution System

Heat is not useful unless it gets to the rooms in the building. The building thermostats sense the temperature in the building, and if the room temperature is lower than the thermostat setting, the heat transport medium (air, water, or steam) is circulated. This reduces the temperature in the heating plant, and at a preset point the fuel burner comes on to raise it again. If the heat transfer medium is not circulating properly, the building may be uncomfortably cold. If it circulates when it's not needed, the building will overheat, wasting fuel.

Any pipes or ducts that run through unheated spaces should be insulated. Any heat lost to such unused spaces is wasted. Other maintenance depends on the building's distribution system.

Steam: Steam is made in a boiler and moves by its own pressure to the radiators. There it condenses back to water, releasing heat to the room. Sometimes the valves of steam radiators allow steam to blow past them so that the water condenses in the return pipes. This wastes heat. Sometimes steam valves fail to shut off and overheating results. Valves on steam radiators should be checked periodically and should be repaired or replaced when necessary.

Water: Hot water is also heated in a boiler, but it must be pumped through the radiators. If a pump is not working properly, one part of the building may be overheating in order to make another part comfortable. This same type of waste can occur when air bubbles block the flow of water through radiators or when valves have been improperly adjusted. Whenever there is uneven heat distribution through the building, pumps and valves should be checked and the radiators bled to eliminate air bubbles.

Air: Air is heated in a furnace and is circulated by fans. A malfunctioning fan or improperly adjusted damper can cause an imbalance in the heat distribution system, making some areas overheat and others be too cold. Whenever this happens, fans and dampers should be checked. Also, air filters within the furnace collect dust and require regular cleaning and replacement.

Air conditioning system maintenance is equally important. Fans, registers, and filters should be cleaned regularly. Exterior, air-cooled evaporators or condenser coils should also be cleaned. Window air conditioners are a major source of heat loss (see Infiltration, page 41); remove them at the end of the cooling season.

What Are the Savings?

It is difficult to estimate precisely the savings due to a properly maintained distribution system. Repairing systems that have chronically overheated a building will result in the greatest savings. The estimated annual savings in the Massachusetts churches was 16 percent to 34 percent ($310 to $3,300).

What Are the Costs?

Costs vary with the type, size, age, and condition of the system; repairs on a badly neglected system could be expensive. Such repairs, however, are basic to any energy conservation program. Once a system is operating properly, its condition can be checked annually. The sexton or maintenance person can do the more routine maintenance tasks, such as bleeding radiators and cleaning air filters, at a minimal cost.

Can It Be Done by Church Volunteers?

Cleaning air filters and bleeding radiators are routine jobs and don't require special training. To save energy this maintenance should be done on a regular basis during the heating season. Volunteers are particularly useful as "watchdogs," seeing to it that any change in the system does not go unattended. The longer a leaking valve or a radiator that won't shut off is left unrepaired, the more heat is wasted.

For More Information

A heating contractor can check the entire system while checking the burner efficiency. It may not be possible to estimate exactly the resulting savings, but the contractor can give estimates on any necessary repairs or service and can also be hired to teach church personnel how to look for and take care of distribution inefficiencies. For larger buildings, consult *How to Save Energy and Cut Costs in Existing Industrial and Commercial Buildings* (see Bibliography).

Rank

Having the distribution system checked at least once a year ranks HIGH. *It is worth doing in every case.*

Most improvements to distribution systems

rank HIGH MIDDLE. The savings are not always worth the expense.

Use It When and Where You Need It: Reducing Demand

Once the efficiency of the heating system has been increased, direct attention to reducing the demand for heat. This section discusses two factors that affect how often and how long the heating plant operates: the amount of space being heated and the temperature to which that space is heated.

It takes more heat to keep a large room warm than a small one. While it is not practical to reduce the size of a building, it is possible to reduce the amount of space being heated at one time. This is especially true when some rooms are used only a few times a week and can be kept at a lower temperature.

It takes much more heat to keep a room at 75° than it does to keep it at 55°. In order for the room temperature to stay constant, the heat flow in must equal the heat flow out. The heat loss from a building is directly related to the temperature difference between inside and outside. For example, if the outdoor temperature is 25°, it will take almost twice as much heat to maintain an indoor temperature of 75° (a 50°-temperature difference) as it will to keep the indoor temperature at 55° (a 30°-temperature difference). To reduce the demand for heat, therefore, heat the least number of rooms and lower temperatures in unused spaces.

The building's thermostats are the keys to reducing demand. Keeping unused rooms at lower temperatures by manually turning down the thermostat settings is the simplest and most straightforward way to reduce fuel consumption. It requires no investment of money, but it does require the education of all building users. Or, for a slight investment, thermostat setbacks can be automated (see page 27).

This same principle applies to service hot water. If water in the hot water tank is kept at 140 degrees during periods when it is not being used, much more heat is lost than if the water thermostat was set at 100 degrees.

Although the following discussion uses the terms heating and lowering thermostat settings, the same principle applies to air conditioning load. By keeping the thermostat at a higher setting the energy demand of the building will be reduced. A word of caution: Be sure you know the kind of air conditioning system the building uses. In what is called a terminal reheat system, more, not less, energy will be used by raising the thermostat.

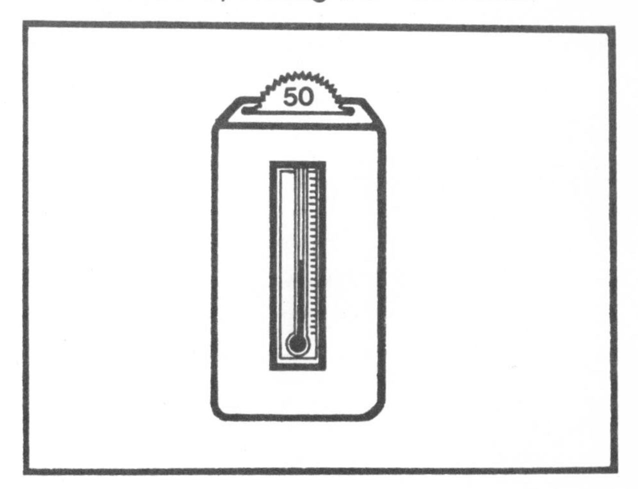

The longer the thermostat setting is closer to the outdoor temperature, the greater the energy savings. In general, this means lowering the thermostat during the heating season and raising it during the cooling season.

This section discusses several methods for reducing demand. First, automatic time clocks can control thermostat settings, hot water temperatures, and excess ventilation. Second, uses can be rescheduled to match heating zones. Third, the distribution system can be rezoned to match actual use patterns. Fourth, ventilation control results in heating or cooling less air.

Automatic Thermostat Setbacks

As we know, keeping the thermostat at a lower setting saves energy. However, manually adjusting the thermostat can mean a cold room and an hour of shivering each morning while waiting for the room to heat. An automatic time clock can eliminate this discomfort. When connected to the building's thermostats, automatic time clocks change the temperature setting automatically. For example, if a room is to be used at 8:00 A.M., the clock can be set for 7:00, allowing an hour for the room to heat. Twenty-four-hour time clocks that can be preset for a daily use pattern are good for rooms such as offices which are used daily. Seven-day timers and clock thermostats can be programmed for a regular weekly schedule and can be changed when the schedule changes. If used properly, they will save their cost in a few months.

When the demand for hot water is intermittent, the service hot water thermostat should be set back. A 100-degree setting will provide enough hot water for occasional use; the lower temperature setting will use less fuel. The temperature can be reset to 140 degrees when more hot water is needed. If the demand for hot water follows a daily or weekly pattern, an automatic time clock should be used.

The building's demand for ventilation is similar to its need for heat. When a space is filled with people or when there is cigarette smoking, for example, there is a demand for fresh air. However, when the building is unoccupied, there is no need for fresh air. Since the ventilation system uses energy, running it wastes heat unnecessarily. If it is inconvenient to regulate the ventilation system manually, a time clock can be used to turn the fan on and off as needed.

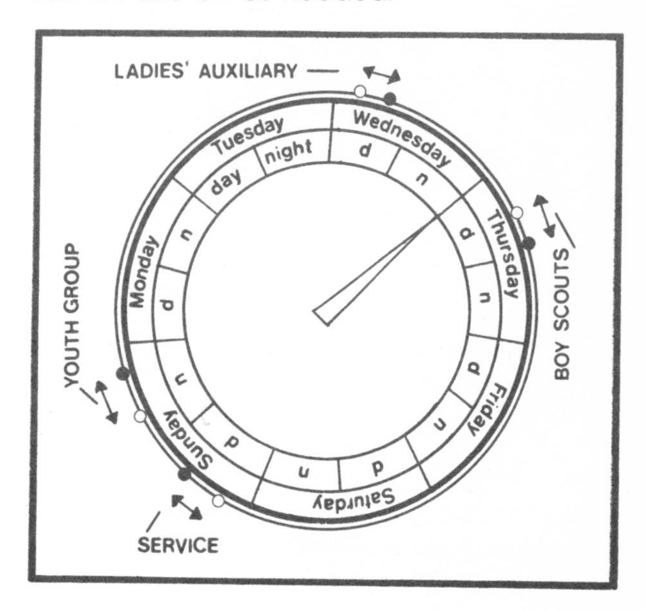

What Are the Savings?

Savings depend on how much and how long thermostat settings and ventilation are reduced; the lower and the longer the setback, the greater the savings. Savings will be greatest in large buildings in cold climates.

What Are the Costs?

Costs include the time clock and its installation. Twenty-four-hour thermostats cost about $50; seven-day clocks cost about $75 plus installation. It may be possible to regulate thermostats, hot water temperatures, and ventilation with a single time clock.

Can It Be Done by Church Volunteers?

Automatic setback thermostats and time clocks should be installed by a heating or electrical contractor. Automation of setbacks and ventilation eliminates the need for voluntary, manual adjustments on a daily basis. In buildings that have many different users, automatic regulation is the only appropriate solution. However, building users will still have to be instructed how to override the preset schedule and how to change the preset schedule to fit new use patterns.

For More Information

By estimating the possible savings, a mechanical engineer may be able to help decide if the cost of time clocks is justified.

Rank

HIGH MIDDLE: Do it if the cost is not substantial. The expense of time clocks will be most obviously justified in large buildings with high fuel costs.

Zones and Scheduling

Different parts of buildings are used at different times. This is especially true of churches; use of the facility may be scattered throughout the week, with the sanctuary being used only on the sabbath. In most large buildings, various zones are controlled by separate thermostats. Users know by experience which thermostat controls each zone. It is important to be aware of these zones when

using thermostat setbacks and scheduling uses of the building.

A plan of the building showing heating zones and thermostat locations is a useful aid to the scheduler. If two rooms are needed, for example, by using ones in the same zone the thermostat can be set back in other zones. Also, activities that use the same space at different times of the day may be scheduled for the same day of the week so that the space doesn't have to be heated as frequently.

What Are the Savings?

Like all setback strategies, the amount of fuel saved depends on how much and how long thermostat settings are lowered. If re-scheduling enables the thermostat settings to be kept lower in larger parts of the building, less fuel will be used.

What Are the Costs?

This option costs nothing to implement, but it does require extra time and attention to zones. If you have a large, complex building, you may want to hire an architect or an engineer to draw up a plan showing the zones and their thermostats; in smaller, simpler buildings this could easily be done by a knowledgeable volunteer.

Can It Be Done by Church Volunteers?

A volunteer may be able to map out the zones. If a copy of this map is posted conspicuously, it can help users to understand how the building operates and help them make intelligent choices about which rooms to use. Education of the users is essential in gaining voluntary cooperation in a setback program.

For More Information

Other than possibly some initial help in drawing up a plan showing how the building is zoned, this energy conservation measure depends entirely on voluntary effort.

Rank

HIGH: Do it without weighing costs and benefits.

Rezoning

The zones in your church may be too large or inappropriate for present use. If, for instance, the entire sanctuary is heated when only the minister's office needs it, an enormous amount of heat is being wasted. If the minister's office could be heated separately, the sanctuary temperature could be lowered. In cases like this, consult a heating or air conditioning contractor to discuss rezoning the building and adding a new thermostat.

There are several changes in zoning that may be appropriate to your building. One is dividing a large zone into two or more smaller ones. Another is changing a room from one zone to another. Another option is appropriate

in some cases: If a very small part of a large zone is used frequently, heating the small space with an electric resistance heater will cost less than heating the entire zone, even with a cheaper fuel.

What Are the Savings?

The benefits of rezoning depend on how large a zone can be controlled separately. The longer and lower the setbacks, the greater the savings. By using an electric heater in the minister's office, for example, the sanctuary temperature could be kept lower. The estimated savings for such a solution in the Massachusetts study was $600 per year.

If rezoning involves changing from heating fuel to electricity, be sure that estimates of possible savings also include the difference in costs for the different types of energy. Electric heat may cost twice as much as heat generated in the building; the savings of heating less space will be partially offset by the extra cost of electric heat.

What Are the Costs?

In some centrally heated buildings, rezoning may be a simple matter of changing a valve or adding a damper. In others it could mean a costly system overhaul. Expense will depend on the type of system and the extent of the changes. The cost of an electric heater for the minister's office is about $60.

Can It Be Done by Church Volunteers?

Rezoning is not a job for unskilled workers, but volunteers can help determine whether or not present zones are practical and useful.

For More Information

A mechanical systems contractor can work up a cost estimate on rezoning. A mechanical engineer or an energy conservation consultant can estimate possible savings based on information provided on the present use of the building.

Rank

LOW MIDDLE: Do it after examining costs and benefits carefully. If benefits are very high, even high costs will be justified.

Ventilation

Mechanical systems in large buildings often provide ventilation in addition to heating. When the ventilation system is operating, a fan draws in cold outdoor air and exhausts warm air. The heat contained in the exhausted air is lost, and more energy must be used to raise the fresh outdoor air to a comfortable temperature.

Fresh air is needed when many people gather in one room and where there is cigarette-smoking or cooking. Ventilation is not necessary when the building is not occupied. Operating the ventilating system when the building is not being used wastes heat. The simplest way to eliminate this waste is to manually turn off the fan whenever the building is not being used. There are three other ways to reduce waste in the ventilation system.

First, it may be possible to reduce the fan's operating rate. This means that less air will be exhausted from the building. A mechanical engineer should be consulted to determine whether or not a reduced ventilation rate would result in discomfort. A mechanical systems contractor would be able to make any necessary adjustment to the fan.

Second, the number of areas that are ventilated could be reduced. Local exhaust fans could be added to areas where there is cooking or smoking. The demand for ventilating the whole building would then be reduced.

Third, the ventilation system could be regulated by an automatic time clock (see Automatic Thermostat Setbacks, page 27).

What Are the Savings?

The possible savings are directly related to the reduction in the ventilation. In heavily ventilated buildings, half the fuel use may be due to the heat lost in the exhausted air. If this loss were cut in half, fuel savings would be 25 percent. Savings from properly controlled ventilation in one Massachusetts church were estimated at 19 percent ($4,000) per year.

What Are the Costs?

No costs are involved when ventilation is reduced by manually turning off the fans. The result is savings in both heating fuel and in the electricity needed to operate the fans.

Reducing the fan rate will involve the preliminary cost of consulting a mechanical engineer and the cost of a mechanical contractor to adjust the fan if it is appropriate.

The expense of adding local exhaust fans can be determined by a contractor.

Placing the ventilation system control on an automatic time clock involves the cost of the time clock and the cost of installation. It may be possible to put both the thermostat setbacks and the ventilation system on the same timer.

Can It Be Done by Church Volunteers?

Turning off the fans, like turning down thermostats, will be most effective in saving energy when all users of the building understand its value and voluntarily do it.

For More Information

Consult a mechanical engineer to evaluate the building's ventilation needs, to estimate costs and savings, and to decide on appropriate strategies. A mechanical contractor can adjust fans and controls and can install automatic time clocks.

Rank

Reducing ventilation by manually turning off the fans ranks HIGH. Consulting a mechanical engineer for a building that has a large ventilating system also ranks HIGH. The rank of the other three strategies will depend on costs and savings as estimated by a mechanical engineer and a mechanical contractor.

Keep It Where You Want It: Reducing Building Load

The previous sections discussed efficient handling of energy within the building. The following pages look at methods for keeping that efficiently produced energy where you want it: in the building.

The basic way to keep indoor temperature different from outdoor temperature is to separate inside from outside by walls, ceilings,

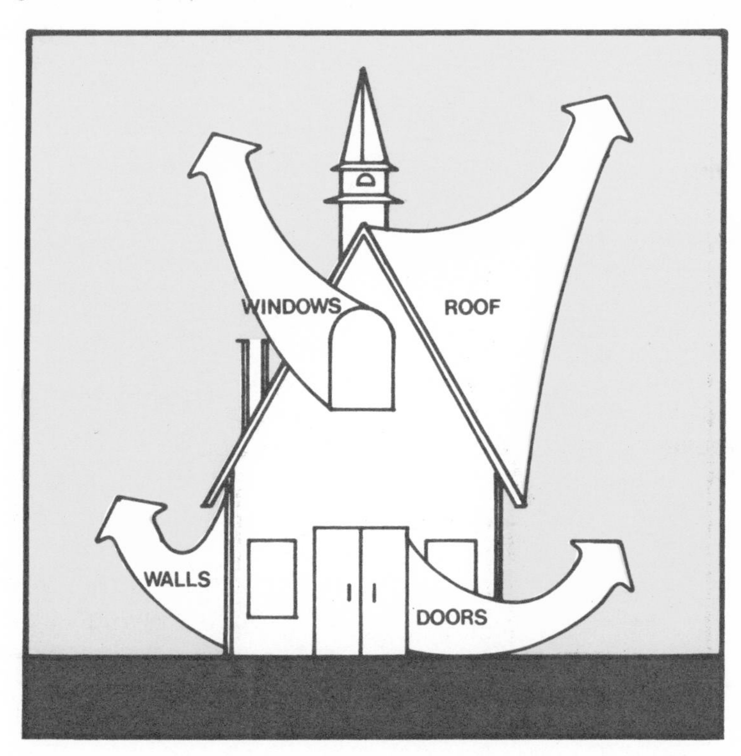

roofs, doors, windows, and floors—that is, the exterior of the building.

The characteristics of these walls, ceilings, and so on determine how efficiently the building holds heat. A building that tends to leak a lot of heat has a *high* heating or cooling "load"; that is, it takes a large amount of energy to heat or cool the building. A building that tends to hold heat well has a *low* heating or cooling load.

Obviously, the bigger the building, the more energy is needed to keep it warm in the winter and cool in the summer. Since it is seldom practical, economical, or aesthetically desirable to reduce the size of an existing building, this measure is not discussed in this book.

The remaining suggestions for reducing building load, taken as a group, are commonly called weatherization and include insulation, weather stripping, and storm windows. These energy conservation measures are more widely known and publicized than the previous suggestions. For church buildings they are less important, but they may still save a significant amount of energy and are worth considering. If the building is air conditioned, these same measures will also reduce the amount of energy required for cooling in the summer at no additional cost. The last three measures of this section are specifically for reducing the air conditioning load.

Insulation

Insulating is the most common and economical method for reducing building load. Insulation is any material that has a high resistance to the flow of heat by conduction. Conduction is the way heat energy moves, by one molecule bumping into another. The more molecules there are bumping into one another, the greater the conduction. For this reason, air is one of the best insulating materials. The trouble with air, however, is that it moves around, taking heat with it. Most good insulating material—such as wool, goose down, or fiberglass—is primarily air with a little bit of something else added to keep the air from moving around.

Heat also moves by radiation. A bright or reflective surface will both absorb and radiate less heat than a dull or dark surface. That is why many kinds of insulation have shiny foil faces. The shiny surface reduces the amount of heat radiating from the insulation through the air.

Types of insulation: Insulation is available in four main forms for different kinds of applications: (1) loose, for blowing or dumping in; (2) batts, for laying in by hand between studs, rafters, or floor joists; (3) rigid boards, for applying over the face of a wall or roof; and (4) liquid foams, for spraying. New insulating material and processes appear on the market each year. Since types of insulation are examined widely in the press, they are discussed only minimally here (see Bibliography).

Installation: The greater the temperature difference between inside and outside, the more heat will flow. For example, insulation installed in a sanctuary that is heated or cooled to comfort temperature only ten hours a week will save less energy per square foot than the same insulation installed in a community hall that must be comfortable twenty hours a week or an office area that must be comfortable forty hours a week. Therefore, install the most insulation in parts of the building that are used the most. This relationship is true for all load-reducing measures; since more energy can be saved in parts of the building that are used the most and therefore heated or cooled the most, put them into practice there first.

If insulation is installed without proper concern for potential moisture buildup, permanent structural damage may result. Airborne moisture tends to condense inside insulation. This condensation diminishes the effectiveness of the insulation considerably. More importantly, it may cause mold, rust, and rot and may shorten the life of the wall, ceiling, or floor.

To prevent this type of damage, install a vapor barrier on the warm side of the insulation. This will keep vapor from entering the insulation and condensing. In a warm, humid climate the vapor barrier should be on the outside of the insulation. In a cool or cold climate the vapor barrier should be on the inside. If in doubt, consult an insulation contractor. The insulation should be as continuous as possible (polyethylene sheets are best) and should overlap at the joints.

A carefully installed vapor barrier will also save energy by reducing infiltration (see section on Infiltration, page 41).

R-values: The ability of insulation to keep heat from leaking out of a building has been

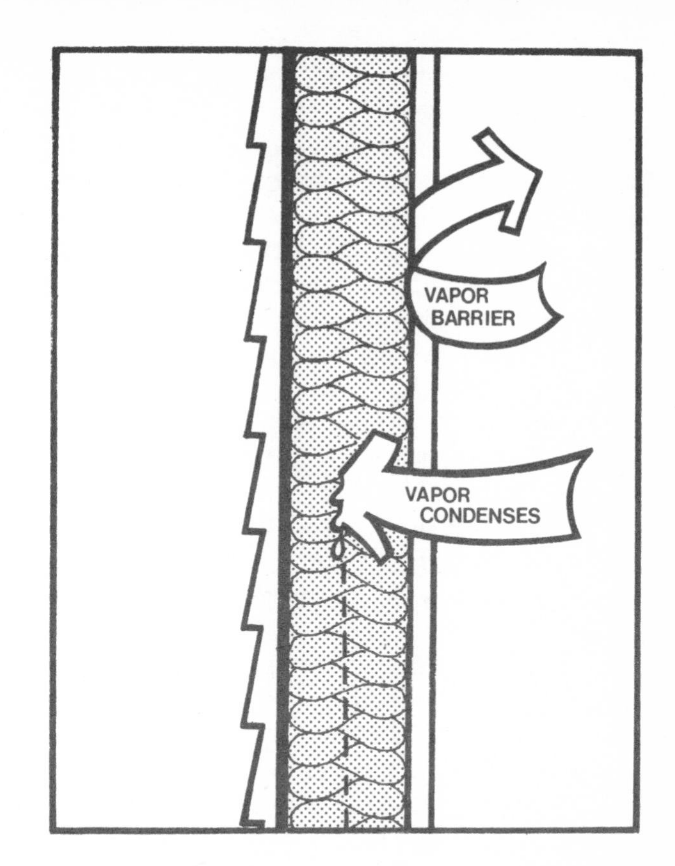

tested and rated. This rating is called the R-value. The higher the R, the greater the insulation's ability to retard the flow of heat and the better the insulation. The R depends on the quality of the material and its thickness. For any given insulation material, the thicker it is, the higher the R will be. An insulation supplier or contractor will be able to indicate the R-value of any insulation under consideration. When shopping for insulation, try to get the highest R for the money.

The relationship between R-values and energy savings is not easy to determine. It involves the R of the existing structure, the R of the added insulation, and their sum—the new R. If the R-value of an uninsulated attic floor is 3 and 3½ inches of fiberglass are added (which has an R of 11), the new R would be 14, and the heat passing through the floor would be cut by $^{11}/_{14}$ or 78 percent. If the same insulation was added to a ceiling that already had an R of 14, the new R would be 25 and the savings would be only $^{11}/_{25}$ or 44 percent.

Two morals to use as guides are:

1. The same insulation will do the most good where the R of the existing structure is lowest. Uninsulated rooms, for example, should be insulated first.
2. Each additional inch of insulation saves less energy than the previous inch. Beyond some point insulation is not saving much energy. Try to establish the optimum level of insulation and stop there.

Costs and savings: The energy savings for insulating a wall, a ceiling, a roof, or a floor depend on several factors: (1) the size of the area to be insulated (the bigger the area, the more energy saved); (2) the R of the existing wall, ceiling, floor, or roof; (3) the R of the insulation; and (4) the temperature difference between the inside and the outside (the greater the temperature difference, the greater the savings). These factors determine how much energy is kept from leaking out of a room. However, this is not the total amount of energy saved; on a seasonal basis, the heating or cooling plant may be operating at 50 percent efficiency, even if properly tuned and maintained. The energy saved in a room should be divided by the average seasonal efficiency of the heating or cooling plant to determine how much *purchased* energy will be saved. These factors should be considered when estimating energy savings from adding insulation.

Costs depend on the volume of insulation installed (that is, the number of square feet times thickness) and will vary from one insulation type to another. Generally, the object is to get the best R per dollar. Be sure to include the cost of vapor barriers in all estimates. Costs also depend on whether or not the installation is in a wall, a ceiling, or a roof.

For cost estimates consult an insulation supplier and a contractor. For savings estimates consult a heating contractor or a mechanical engineer who is familiar with different methods of heating and cooling buildings. A church member may also be able to prepare an estimate, using procedures outlined in *Other Homes and Garbage* or *ASHRAE Fundamentals* (see Bibliography).

For a general discussion of insulation, see February 1978 *Consumer Reports* and *Insulation Manual.*

Ceiling Insulation

Ceiling insulation is installed in the space above the ceiling and below the roof (the attic). Generally, it is installed between the floor joists of the attic floor, with a vapor barrier placed on the warm side of the insulation. The space above the insulation should be adequately ventilated. Natural ventilation through vents in the roof eaves or soffits and louvers in the gable end walls is sufficient and will prevent heat or moisture buildup in the attic space. If insulation is of the loose or blown variety, be sure it does not clog these vents.

The most popular types of insulation for ceilings are fiberglass batts, mineral wool, and cellulose. If the roof of the attic is low, insulation may have to be blown in; this should be done by a good insulation contractor.

What Are the Savings?

Each of the three Massachusetts churches had a room or rooms where ceiling insulation could be installed. Estimated annual savings were 3 percent to 11 percent ($42 to $1,400).

What Are the Costs?

Costs vary, but insulating a ceiling generally costs less than other insulation measures because there is usually no finish material or touch-up required. Also, there is usually enough space to install the desired amount of R with the least expensive insulation. In the Massachusetts churches, cost estimates varied from $100 to $3,000.

Can It Be Done by Church Volunteers?

If there is enough working height in the attic and the ceiling is adequately supported, installation of ceiling insulation is an excellent project for volunteer labor.

For More Information

Installation techniques are discussed in *In the Bank . . . or Up the Chimney?* (see Bibliography). For savings estimates, consult a heating contractor or a mechanical engineer.

Rank

HIGH MIDDLE: Do it if the cost is not substantial. Costs may be high, but benefits are often equally high. Analyze carefully.

Roof Insulation

Roof insulation can be applied to the top of the roof or to the bottom of the roof. Insulation applied to the bottom of the roof is essentially the same as wall insulation (see page 39). When applying insulation to the bottom of the roof, a ventilated air space should be left above the insulation and should be vented top and bottom to the outdoors to prevent moisture buildup and to help prevent roof leaks.

Rigid fiberglass or foam board is applied to the top of the roof. In some cases it may be installed over the existing roofing. Sometimes, however, the roofing will have to be removed. Consult an architect or top-quality roofing contractor. In either case, new roofing will need to be installed. Since the roofing material is more expensive than the insulation, the best time to apply roof insulation is when you are planning to reroof.

Roof insulation is generally limited to a 2-inch to 3-inch depth. If more is installed, the roofing cannot be attached properly to the structure. Because depth is the limiting factor, you may want to use urethane to get the best R-value per inch. Or you may select a roofing system that can tolerate a thick layer of insulation. For instance, on flat roofs an extra-heavy layer of stones can be used to hold down a vinyl roof without any attachment through the insulation.

What Are the Savings?

The savings depend on the size of the roof, the amount of space below that is heated or cooled, and the R of the existing structure. Estimated savings in one Massachusetts church were 13 percent ($170) per year.

What Are the Costs?

Since roof insulation must always be combined with reroofing, it is expensive. However, if the building is being reroofed anyway, the additional cost of insulation is small. Costs in one of the Massachusetts churches were estimated at $250, not including the cost of reroofing.

Can It Be Done by Church Volunteers?

Some roofing projects are suited for volunteer workers, others are not. Insulating a shallow-pitched roof with asphalt shingles or roll roofing (especially if close to the ground) is an acceptable project for volunteers. Most of the work requires only a hammer and good balance. However, close supervision by an experienced person is essential in order to avoid leaks and to achieve a good finished appearance. Volunteer roofers make it economical to install roof insulation even when the church would not otherwise reroof. Insulation on flat roofs and on extremely steep roofs should be installed by professionals.

For More Information

The best sources on reroofing and insulating techniques are experienced building professionals, such as architects and top-quality roofing contractors. For an estimate on energy savings and advice concerning how much insulation to install, consult a mechanical engineer or a heating contractor. A general building contractor will be able to advise on techniques and costs for a small, simple-shingled roof.

Rank

LOW: Do it only after other energy conservation measures have been implemented.

If the plan is to reroof the building anyway, then roof insulation should be ranked LOW MIDDLE. The cost may be substantial, but the benefits are usually worth it. Analyze carefully to determine the best R for the money.

Wall Insulation

Churches built before 1960 probably have uninsulated walls (others may also). Uninsulated walls fall into one of two categories: hollow walls or solid walls.

Hollow walls are usually composed of wood or metal studs with a finish material on either side, forming a cavity where insulation can be installed. The materials available and the R-values for filling a 3½-inch cavity are as follows*:

Insulation	*R-value*
Cellulose fiber	11.95
Urea-formaldehyde foam	16.50
Vermiculite	6.90

Hollow walls have no vapor barrier, and since there is no way of installing one, this presents a major problem. Cellulose fiber and loose vermiculite allow the passage of moisture. Urea-formaldehyde foam is generally impervious to water vapor, but it shrinks and cracks where it contacts other materials. At these places the other materials will rot or rust if they become moist. Because there is little information presently available on the effect of adding insulation to an existing wall without

*These figures represent standard insulation values minus 1.0, the R-value of the air space lost by filling the void with insulation.

a vapor barrier, insulating under these conditions is generally not recommended. Wait until better information or better insulations are available. However, if the building is usually dry and if the area where the wall is located is not exposed to moisture from such things as cooking, the saved energy may be worth the risk.

Other alternatives are possible. Covering the inside surface of the wall with a vapor-proof paint may achieve a reasonable vapor barrier. If moisture levels are low, this will probably be satisfactory, but the paint may have to be reapplied periodically. Or rigid insulation may be added to the face of the present wall, as if the wall were solid.

Solid walls are usually of masonry. Insulation is applied to either the interior or the exterior face of the wall and is then covered with a new finish material. Use rigid board insulation (fiberglass or foam) or add new studs or furring strips with insulation between the studs (fiberglass batts are usually the best buy). In either case, there should be a vapor barrier on the warm side of the insulation.

Should the insulation be installed on the inside or on the outside? Normally, it is better to put the insulation on the inside of the wall; interior finish materials do not have to be weathertight, as they would on the exterior, and thus they cost less. However, for churches in a climate where the air conditioning expense is higher than the heating expense, it can be advantageous to put the insulation on the outside of masonry walls in the parts of the building that are air conditioned for forty or more hours per week. In this way the masonry inside the insulation will store coolness. Air conditioning requirements will be reduced at the beginning and at the end of the air conditioning season and on days when the daily average temperature is comfortable. This stored coolness also permits air conditioning to be turned off during brownout (peak demand) periods without having uncomfortable temperature rises.

What Are the Savings?

For factors influencing how much energy will be saved, see the section on Insulation. Two of the Massachusetts churches showed a possible annual savings from 6 percent to 18 percent ($94 to $185).

What Are the Costs?

These factors will influence costs: (1) the size of the wall, (2) the kind of material used as insulation, and (3) whether or not the wall is easily accessible from the ground or from indoors. Insulating a solid wall generally costs more than insulating a hollow or cavity wall because (1) rigid insulation costs more per R than loose insulation, (2) a new interior or exterior finish is required, and (3) new trim must be added around doors and windows due to the extra thickness of the wall. Estimated costs for two of the Massachusetts churches were $770 and $4,400.

Can It Be Done by Church Volunteers?

Insulating cavity walls with blown-in insulation is not a job for volunteers. On the other hand, adding insulation to an exposed stud wall or to the inside face of a solid wall by furring or studding and adding drywall is an excellent volunteer project, but it requires good supervision. A professional may be needed to install new trim around windows and doors.

For More Information

Check the sources under *Insulation* in the Bibliography.

Rank

LOW MIDDLE: Do it after estimating costs and benefits carefully. Costs are substantial, but benefits may be high also.

Infiltration

Although insulation slows heat loss, up to 50 percent of a building's heating load is due to air leakage through the exterior. Air that leaks into the building must be heated or cooled to room temperature. The same amount of already heated or cooled air escapes, wasting the energy it took to bring that air to room temperature. This leakage around doors and windows and through small cracks and open doors or windows is called infiltration.

The amount of infiltration depends on the length and width of the cracks, how often doors and windows are opened, and the general tightness of construction. Doors and windows that are often open, numerous cracks, long or wide cracks, a generally "loose" construction, and high wind speeds greatly increase infiltration.

The temperature difference between the inside and the outside influences the amount of heat lost or gained by infiltration. The greater the temperature difference, the greater the amount of heat passing through the wall. As discussed in Automatic Thermostat Setbacks (page 27), the temperature difference will be larger when rooms are heated or cooled to comfortable temperatures than when they are not. Therefore, infiltration carries more heat through the walls of the rooms that are used the most.

It is impossible (and undesirable) to eliminate this air leakage entirely, but it can be cut substantially by installing weather stripping, caulking, storm windows and doors, shutters, thermal curtains, and plantings.

Weather Stripping

Operable windows and doors have cracks around their edges when closed. Weather stripping material fills these cracks; when a door or window shuts, it compresses to fit snugly and stops the flow of air. Weather stripping, therefore, saves energy by stopping warm air from going out during the heating season and from coming in during the cooling season. It also eliminates drafts and allows the thermostat to be lowered.

There are several kinds of weather stripping. Metal v-strips are suited for the edges of doors and some windows and last longer than other types, such as tubular gaskets, reinforced felt, foam-edged wood, and foam

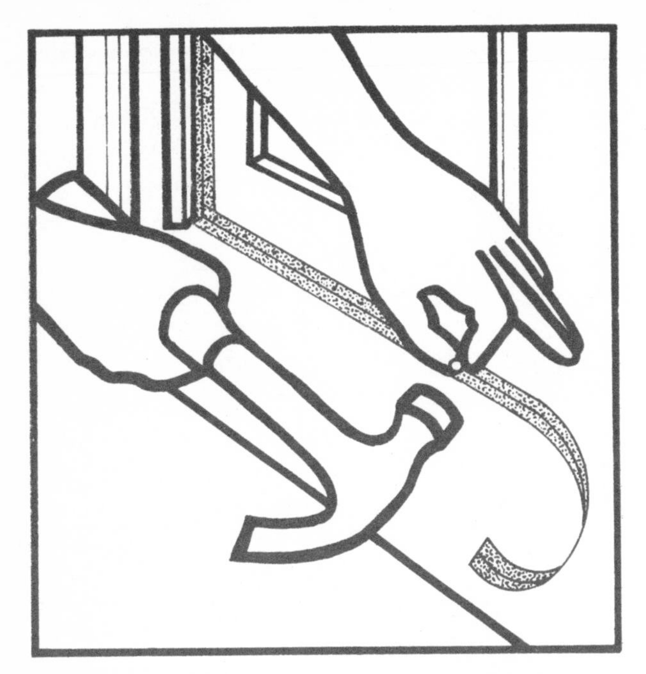

tape. In addition, there are special strips for the bottoms of doors. Weather stripping should be replaced every two to five years, depending on the type and extent of wear. The usual signs of wear are obvious crimps, bends, and loosening. Rope caulk is the least expensive but lasts only one year.

What Are the Savings?

Energy savings depend on the size of the cracks; the bigger and the longer the cracks, the more energy weather stripping will save. Weather-stripping old, leaky windows and doors and windows that do not have storm sashes will save the most energy, especially where the space is most often heated (or cooled) to room temperature.

What Are the Costs?

The expense of weather stripping is determined by the type used. The better kinds cost two to three times as much but may last two to three times as long as cheaper products. Installation cost is a major factor but usually does not vary much from one type to another. As a rule, buy a durable, better quality weather stripping.

Can It Be Done by Church Volunteers?

Weather-stripping is an excellent project for volunteers. It can be done a little at a time or by one large work party. It requires little skill and no exotic tools. A word of caution. Improperly installed weather stripping will not last. Loose weather stripping will not only be ineffective at stopping infiltration; it will also be an eyesore. Be sure to have someone who is careful and knowledgeable on hand, and put special emphasis on doors and windows that are opened most frequently.

For More Information

Contact a local weather stripping supplier, hardware store, building supply outlet, or local contractor regarding cost information and advice on the best type of weather stripping for your particular installation. For installation details see "Weatherstripping," *Consumer Reports,* February 1977. Estimating the energy savings is difficult and is usually not worth the effort. If you are interested, however, estimating techniques can be found in *Project*

Retro-Tech and *Other Homes and Garbage* (see Bibliography) or consult a mechanical engineer or a heating contractor.

Rank

LOW: Do it after everything else is done. Costs are not high, but neither are the benefits in most churches. If volunteer labor is used, rank this measure LOW MIDDLE: Costs are low and benefits are moderate. However, unless the building is subject to an overall energy audit, it usually costs less to install weather stripping than to do a detailed estimate of savings. If in doubt, spend money and time on weather-stripping the most obvious cracks rather than on weighing costs and benefits.

Caulking

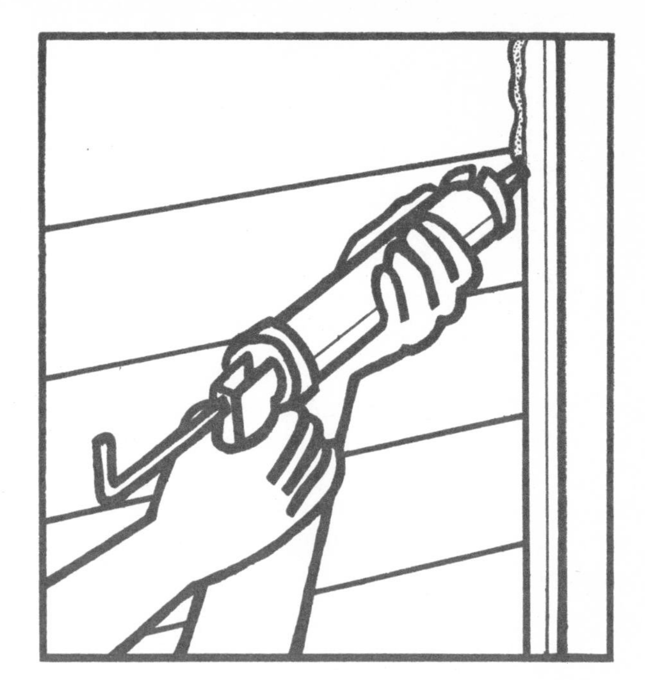

Whenever two hard materials are joined together, a crack forms at the joint. Changes in temperature and humidity open these cracks, and air leaks in and out. Caulking is a soft, flexible, sticky material that sticks to both sides of the crack. Since it stretches and compresses as the crack opens and closes, the crack is always filled, completely stopping the air flow.

Caulking, like weather stripping, comes in many forms, from ropes and knife-grade for filling wide cracks to gun-grade cartridges and tubes for smaller cracks. There is a wide variety of chemical compositions. Their performance, durability, and cost vary, so choose carefully.

Cracks are most likely to appear where a wooden door or window frame meets a masonry wall, where siding abuts a masonry chimney, or at masonry sills (along the top of the foundation). Structural shrinkage or settlement can also cause cracks. Caulk around flanges, air conditioners, storm windows, and any pipes or ducts passing through the walls too.

What Are the Savings?

The wider the crack, the more air leaks in, and, therefore, the more energy is saved by caulking. Also, caulking around spaces normally heated or cooled to room temperature saves more than caulking around unused spaces. In addition to saving energy, caulking will help keep out rain, insects, and mold and will lessen structural deterioration. Some caulkings deteriorate within a year, while others will last twenty years and therefore will save twenty times as much energy. Savings for caulking and weather stripping in one of the Massachusetts churches were estimated at 5 percent ($470) per year.

What Are the Costs?

Costs depend on the quality and durability of the caulking. Since better-quality caulking may last much longer than a less expensive medium or poor-quality caulking, it is generally worth the extra expense. Of course, the wider and the longer the cracks, the more caulking will be needed. In the three Massachusetts churches, for example, costs for caulking and weather stripping, using no volunteer labor, were estimated at $2,500.

Can It Be Done by Church Volunteers?

Caulking is an excellent volunteer project, requiring little skill and inexpensive tools, such as a caulking gun and a putty knife. Of course, at least one knowledgeable person should choose the caulking materials and coach the installation. Some caulking materials stain badly and some, or their solvents, can be toxic or irritating; use caution.

For More Information

For more information on types, durability, and special uses, see the Bibliography. For cost information consult a local hardware store, building supplier, or contractor. As with weather stripping, estimates of energy savings are inexact. Methods for estimating are described in *Other Homes and Garbage* and *Project Retro-Tech* (see Bibliography), or consult a heating contractor or a mechanical engineer.

Rank

LOW: In most churches costs are high in relation to benefits. However, if volunteer labor is used, costs are low. Benefits are still moderate, but they vary widely and may be substantial. Caulking wide, obvious cracks ranks HIGH MIDDLE: It is worth doing as long as costs are not substantial.

Storm Windows and Doors

Insulation reduces conduction; caulking and weather stripping reduce infiltration; storm windows and doors reduce both infiltration and conduction. Conduction, for example, is reduced by trapping a layer of air between the storm door or window and the regular door or window. Combination aluminum screen/storm doors and windows, for instance, add an R-value of about 1.4. This represents a savings of from 40 percent to 60 percent of conduction losses through the window. Storm windows and doors also lower infiltration by providing another barrier to the outside air. This saves 30 percent to 50 percent of infiltration losses through the window.

In northern climates, if storm windows or doors are installed on the outside as is usual, they must be vented slightly to the outside to prevent moisture buildup and condensation. If the existing window is very leaky, it is better to install storm windows on the inside and to make them tight. Otherwise, condensation is likely to occur on the storm window constantly.

What Are the Savings?

The more windows and doors that are equipped with storms, the more energy will be saved. Storm windows and doors save more energy in spaces that are normally heated or cooled to room temperature, so equip these first. Storm windows and doors can only stop the heat that is presently leaking through the existing windows and doors. Therefore, as you would expect, storm windows and doors save the most energy where existing windows or doors are the leakiest. They save less over double-paned windows than over single-paned windows. Wooden storm windows save 10 percent to 20 percent more energy than aluminum ones, because the wooden frame is a better conduction stopper. By adding storm windows to unprotected windows, one Massachusetts church was estimated to save about $230 per year.

What Are the Costs?

Costs vary. Homemade, wooden-framed, plastic storm windows are short-lived and are usually unacceptable aesthetically; but they are cheap. Standard aluminum storm windows and doors are not as effective and they cost more, but they are durable and require low maintenance. Wooden storm windows and doors cost the most and require maintenance. A major factor in the cost can be whether or not the window is a standard size or if the storm window must be custom-fabricated. Windows with elaborate tracery and stone frames are expensive. However, these elaborate windows often use valuable stained glass. Insurance rates can sometimes be reduced if these stained-glass windows are protected by plastic glazing. The protective glazing can be converted to a storm window by making the edges tight.

Storm doors on public buildings, such as churches, may constitute a fire hazard by blocking emergency exits. In general, if the existing door swings out toward the exterior, the storm door will need to swing in, and thus creates a safety hazard. In such cases, a vestibule can be built between the interior and the exterior doors. This would allow both doors to swing in the direction of emergency travel. However, this adds substantially to the cost.

Costs for tightening the protective glazing in one large Massachusetts church were estimated at $1,200, with savings of $831 per year. The estimated costs for adding storm windows to unprotected windows were between $6,000 and $12,000.

Can It Be Done by Church Volunteers?

Constructing wooden-framed, plastic storm windows is an acceptable task for church volunteers and is a good self-help project. Both aluminum and wooden storm windows and doors can be purchased from a hardware supplier and can be installed by volunteers under competent supervision. In addition, a wooden storm window applied over an operable sash will usually need to be removed each spring and to be rehung in the fall. This too can be done by volunteers. Of course, many operable windows in churches are never opened, and their storm windows may be left on year-round.

For More Information

For information about costs consult a local hardware store, building contractor, or storm window installer.

Savings are difficult to estimate. Check the resources listed in the Bibliography. A heating contractor or a mechanical engineer can supply a realistic estimate of the amount of heat that can be saved.

Rank

LOW: Do it after everything else is done. Costs may be substantial in relation to savings.

Shutters

Shutters, like storm windows and doors, slow down both conduction and infiltration. They also significantly lessen radiation. This is particularly important in air conditioned buildings where the cooling load is increased by radiant sunlight.

There are several kinds of shutters: interior, exterior, sliding, swinging, bifold, fixed, and "soft." To be effective a shutter should be tight and weather-stripped at all edges and should be made primarily of an insulating material such as foam. Although a way may be devised to operate exterior shutters from the inside, interior shutters are easier to operate because they are more accessible. Be sure the materials used for interior shutters are acceptable under local building codes and do not represent a fire hazard. The local building official can help determine this.

What Are the Savings?

The more windows that are shuttered and the longer they are closed, the more energy will be saved. As with storm windows, savings are greater when shutters are installed in rooms that are used the most and on leaky, metal, leaded, or single-glazed windows. The tighter the shutter and the greater its R-value, the more energy is saved. Since shutters only save energy when they are closed tightly, put them where they can be closed most of the time, for example, in meeting rooms that are used only two or three times a week but are heated every day because they are in the

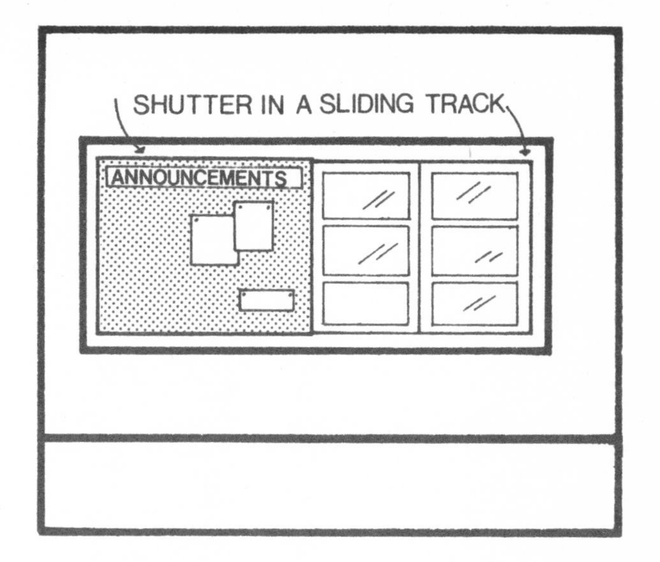

same zone as offices. On a window facing south, open the shutter daily during the heating season to let in the sunlight. Estimated annual savings in the Massachusetts church study were 6 percent ($36).

What Are the Costs?

Shutters can be bought or built for a wide range of costs, depending on their construction. Interior shutters generally cost less than exterior ones, because they do not have to be built to withstand winds, rain, hail, and so on. A simple shutter is a piece of foam board fitted snugly into the window frame, with weather stripping at the edges. More complex versions have special decorative coverings, rigid frames, sliding tracks, locking hardware, and operating mechanisms. These increase the cost. Decorative exterior shutters add to the appearance of a building. With a little thought and small cost, they can often be transformed into thermal shutters. Even if they cannot be made to fit tightly, exterior shutters should be closed during the air conditioning season to reduce solar heat gain. If interior shutters can be used as bulletin boards or as room-darkeners, their cost is easier to justify. Estimated costs for interior shutters in one Massachusetts church sanctuary were $500.

Can It Be Done by Church Volunteers?

Making shutters is a good project for volunteers who have the necessary skill and tools. Making decorative shutters can be incorporated into the Sunday school or youth group programs. A "soft" shutter made of upholstery foam and covered with fabric would be an excellent project for a needle-point guild or sewing group.

For More Information

Thermal shutters will soon be available from manufacturers on a commercial basis. However, most shutters must be customized to fit particular windows. A number of designs have been published in such sources as *Mother Earth News* and *Popular Mechanics,* and more are rapidly becoming available. Consult local artisans, carpenters, and contractors.

Rank

LOW: Do it after everything else is done. Costs are usually high in relation to savings. However, special circumstances that reduce costs (such as volunteer labor or using existing shutters) or increase benefits (such as a far north location or multiple use) justify the expense. Also, because of the "volunteer project" nature of this measure, it can be an excellent method for increasing conservation awareness among church members.

Thermal Curtains

Curtains, like shutters, are energy-savers in three ways: they reduce conduction, infiltration, and radiation. Their appearance, again like shutters, can be varied to suit particular needs. They can be roll-up shades, draw curtains, or sheets of cloth stretched over a frame. As with shutters, seal them tightly at all edges and use an insulating material such as quilting fabric. Made-to-order curtains are being introduced on the market. Or, invent your own. As with shutters, avoid creating a possible fire hazard by making sure the materials are acceptable under local building codes. The product manufacturer can supply fire hazard classifications, flame spread ratings, and so on. Also, be sure curtains do not block air flow around heating registers and radiators.

What Are the Savings?

The more windows with curtains and the more they are closed, the more energy will be saved. The most energy is saved in rooms where the curtains can be closed when the room must be heated or cooled, for instance,

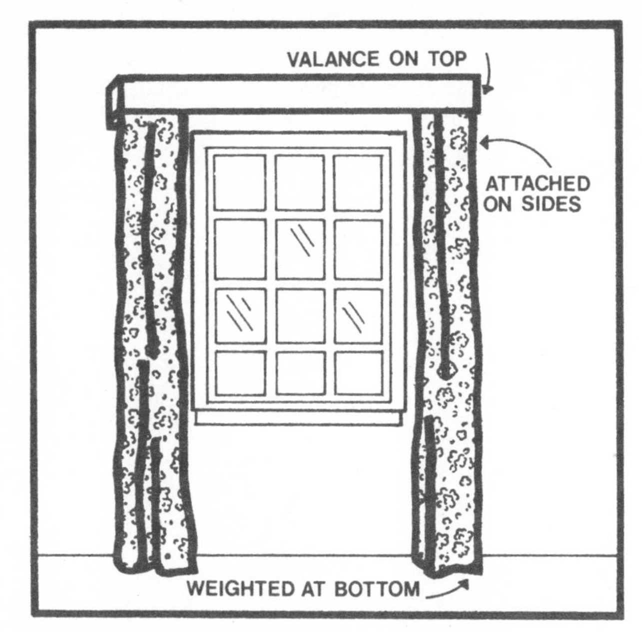

rooms used primarily at night. They will also save the most on leaky, metal, leaded, or single-glazed windows, since these windows lose the most heat. The tighter the edges and the greater the R-value of the curtain the more energy will be saved. If curtains on south-facing windows can be opened daily during the heating season, solar heat will be gained. Generally, curtains can be expected to save as much as shutters.

What Are the Costs?

Curtains can be bought or made for a wide range of costs. When comparing costs of made-to-order curtains, be sure to compare them equally. For example, one curtain manufacturer may quote costs for the curtain including installation, while another may quote only the cost of the curtain. Also, be sure to add any special requirements such as installation into masonry or extra long cords and pulleys for high windows.

If new curtains are needed anyway, thermal curtains will cost little more than conventional ones. In 1978 thermal curtains cost about $2.50 to $3 per square foot, excluding shipping or installation.

Can It Be Done by Church Volunteers?

Making thermal curtains is a good project for volunteers. A sewing group or needlepoint guild may find them to be just the right combination of decoration and function. If you do make your own, look carefully at several made-to-order designs to get ideas and improve on their weaknesses. Care and skill in installation are required to avoid leaky edges.

For More Information

Ask manufacturers of made-to-order thermal curtains for their literature, and actually examine their product if possible. These curtains can usually be found at energy fairs and displays and are advertised in magazines that encourage energy conservation. Designs for thermal curtains appear in such publications as *Mother Earth News* and *Popular Mechanics.* A local curtain store that features custom-made draperies may be of some help too.

Rank

LOW: Thermal curtains normally cost more than they save. Special circumstances that reduce costs (such as volunteer labor or the fact that regular curtains would be purchased anyway) or increase benefits (such as a far north location or use as a room-darkener for audiovisual projection) may justify the expense.

Vegetation

Strange as it seems, trees and bushes lower energy consumption. Trees and bushes growing close to a building will reduce the wind speed at the building surface by up to one third. This reduces the pressure at which the wind pushes outside air into the building and, therefore, reduces infiltration.

In climates where heating costs are significant, plantings should be evergreens so that they will be able to slow down the wind during the winter. They should be planted on the side from which the prevailing winter winds blow (check with a local weather station). Place them so that they flank the window or door but do not block the view.

What Are the Savings?

Evergreens planted next to windows and doors will save the most energy. They will do the most good near the leakiest windows and doors and near rooms that are normally heated or cooled to room temperature. They will save less where storm windows or doors, thermal curtains, or shutters are already in use. In one Massachusetts church, planting ten evergreen trees was estimated to save 3 percent ($30) in 1977.

What Are the Costs?

Costs of evergreens vary widely with type and location. A lot will depend on the climate. Small trees and shrubs are particularly suitable for planting next to basement windows and are less expensive than larger varieties. Someone may be willing to donate evergreens or a local nursery may give a discount. Often people who want to thin the shrubbery around their houses are glad to donate the surplus. If someone wishes to donate some plantings or if the church is planning a landscaping program, planting evergreens on the windward side of the building may not cost anything extra. In the church mentioned above, the cost of ten evergreens was estimated at $750 in 1977.

Can It Be Done by Church Volunteers?

Planting evergreens is a good volunteer project. It requires a little knowledge, less skill, and is good exercise. Volunteers can help dig up a church member's overgrown shrubs, transplant them at the church, or buy and plant nursery trees. Be sure the shrubs and trees receive proper care after transplanting or the investment will turn into an eyesore.

For More Information

Consult a local nursery for information on the best type of tree or shrub for your church. If you are doing a lot of planting, consult a landscape architect. However, do not expect any of these sources to be able to give an accurate estimate of how much energy the plantings will save.

Rank

LOW: Benefits are small but enduring. Costs may be substantial unless you plan to plant anyway or use volunteer labor to plant donated trees or shrubs.

Keep Out the Summer Sun: Reducing Solar Heat Gain

The previous measures suggested for reducing building load will effectively reduce both heating and air conditioning. The suggestions in this last section are for reducing air conditioning energy only. In fact, if not done properly, they could increase the energy required for heating. Each of them reduces the amount of direct sunlight that comes through the windows and heats the building. During the winter the sunlight provides desirable heat energy. During the summer its heat must often be removed by air conditioning. When using these measures, be sure the heating load is not increased more than the cooling load is decreased. Concentrate on decreasing heat gain through east or west windows, but shade southern windows from overhead summer sunshine only. Let the low winter sun still shine in the southern windows.

Shading with Vegetation

Shade trees planted on the west, south, and east sides of the building decrease the amount of sunlight coming in the windows and heating the roof and walls. If your church is in a climate where heating bills are significant, plant deciduous (leaf-shedding) trees. They will allow the solar energy to heat the building in the winter when needed, while providing shade during the hot summer months. If your church is in a climate where you usually must air condition, evergreens will give year-round shading and will often grow faster.

For the south side of the church, try to get tall trees with branches that reach out over the roof. Since branches are not needed at window level, trees should be planted as close as possible to the building.

On the east and west, plant trees with low branches. These lower branches will protect the east and west windows and walls from the blast of summer sunshine in the morning and in the afternoon. Tall trees, if located close to the building on the east and west, will help shade the roof. However, it is most important to shade the west windows against the afternoon sun. Try to plant so that the sun is blocked when it is about halfway between noon and sundown on a summer day. Often this can be done by planting medium-sized trees a slight distance away from the building; if they are growing along a nearby property line, for instance, they block the sun but not the view.

What Are the Savings?

The more shading provided during the cooling season, the more energy will be saved. Shade particularly those portions of the building used (and therefore air conditioned) most, such as offices. If the windows have reflective or absorptive glass, trees will save energy but not as much. The denser the foliage and the bigger the tree, the more it will save.

What Are the Costs?

Large trees cost more than small ones, but all trees vary widely in cost, depending on location and type. Shop around. If the cost of large shade trees is prohibitive, settle for a sapling that will grow larger. Or transplant a small tree from an unwanted place to where it will do more good.

Can It Be Done by Church Volunteers?

A fifteen- to twenty-foot tree is too large to handle without special equipment, but plant-

ing small trees is a good project for volunteers. Little special knowledge or skill is required, and the benefits are fresh air and exercise. Anyone can participate. It can be an especially attractive youth group project if the group also takes responsibility for caring for the tree after planting.

For More Information

Consult a local nursery for information on types, care, costs, and transplanting. If extensive planting is planned, consult a landscape architect to get the most for the money aesthetically, as well as in energy savings. A mechanical engineer who is experienced in building design will be able to estimate the energy savings.

Rank

HIGH MIDDLE: Costs may be substantial, but benefits may be also. Benefits will be proportional to the amount of air conditioning used now. For example, if the building needs to be air conditioned forty hours per week, you will save more than if you air condition twenty hours per week.

Shading with Structure

A number of devices will prevent the sun from directly striking windows. These devices may be placed on either the inside or the outside of the window.

Outside shading will cut solar gain through windows by as much as 80 percent. Sunlight is most intense in the middle of the day. However, at noon in the summer, when the sun is high in the southern sky, it strikes the windows at a steep angle and is mostly reflected back out. The solar heat from the lower eastern and western sun is much greater, because it shines more directly into the window. Louvered screens will shade low eastern and western sun while still allowing indirect, sky light to enter and light the building. These screens will, of course, block some of the view. A surprisingly small roof overhang or overhead sunscreen will shade southern windows from the high southern sun. This device will not block the view.

Blinds and screens on the inside of windows are less effective than on the exterior, because the sun has already entered the building by the time it strikes the shade. A reflective surface on the outside of the shade

will greatly increase its effectiveness by "bouncing" the solar radiation back out of the building.

What Are the Savings?

Solar heat gain tends to be most critical through western windows, because the afternoon air temperature is usually higher and air conditioning is more likely to be turned on. Therefore, savings are usually greatest from shading western windows. Exterior sunscreens or interior blinds on east and west windows will reduce the cooling load more than on the south. The longer the cooling season, the greater the savings. Shading will best benefit buildings with large glass areas and buildings where the climate is clear and dry. Here the air conditioning load is due primarily to direct solar radiation. On the other hand, in buildings that gain heat because of high air temperatures and that are in humid and overcast climates, shading will affect a smaller portion of the air conditioning load.

What Are the Costs?

Because exterior sunscreens alter the appearance of the building, they should be attractive. The time and care this requires will tend to make them expensive. Expect to pay at least $5 per square foot (in 1977) for a wooden screen. Also, because sun angles vary at different latitudes and at different building orientations, sunscreens should be custom designed or their design should be reviewed by an architect. Since these costs are high, interior shades, although less effective, may be the only feasible option. A reflective surface on the outside of these interior shades is essential.

Can It Be Done by Church Volunteers?

The screens or shades can be assembled and erected by volunteers with carpentry experience under the supervision of a knowledgeable person. This will substantially reduce their cost. Also, sanding and painting are jobs that take time but require little experience. Volunteers could also help with the installation of interior blinds or shades.

For More Information

The possible savings can be calculated by a mechanical engineer. Exterior screens should be custom designed or their design should be reviewed by an architect or an engineer experienced in shading devices. Once a design is established, a building contractor or carpenter can make the screens and install them. Secure information on conventional awnings and interior shades from builders' supply outlets, hardware stores, and interior decoration outlets.

Rank

LOW MIDDLE: Savings can be high, but so can costs.

Tinted or Reflective Glass and Film

If it is not feasible to shade windows, it is still possible to reduce cooling needs by adding reflective film to clear window glass and by replacing clear glass with tinted or reflective glass. Normal window glass transmits from 85 to 90 percent of the solar radiation that strikes it; bronze, gray, or green tinted glass absorbs radiation and transmits only half as much into the building. Because tinted glass heats and expands when it absorbs radiation, it should not be shaded. Stresses from differential expansion can crack it.

Reflective glass and reflective film on clear glass cut heat gain by reflecting rather than by absorbing light. They have a mirrorlike appearance on the outside during the day and on the inside at night, when the lights are on. Adding film to clear glass should be done carefully to prevent distortion of the view.

What Are the Savings?

Tinted or reflective glass will be most beneficial in buildings that have high cooling

loads. They will cut the cooling load most effectively in hot, dry climates, where a larger percentage of unwanted heat is from direct sunlight. If reflective or tinted glass is used in windows that have been contributing heat in the winter, the cost of heating the building will rise. For this reason, use tinted or reflective glass and film in buildings where the cost of cooling is higher than the cost of heating, or use it on east and west windows only, not south.

What Are the Costs?

This is a fairly costly way to reduce heat gain. For example, in January 1977 the average cost of clear plate glass was 85 cents per square foot. Tinted glass was $1.10 per square foot, and reflective glass was $1.90 per square foot. It is easier to justify using tinted or reflective glass in new construction, where only the difference in cost between it and ordinary glass is attributed to energy conservation. In existing buildings the cost of replacing window glass is much greater.

Reflective film costs much less but may be aesthetically objectionable unless well installed.

Can It Be Done by Church Volunteers?

Replacing window glass or adding plastic film to glass is not an appropriate job for unskilled volunteers. However, an experienced, skilled do-it-yourselfer may be able to do a good job if the panes are not too large.

For More Information

A building contractor or an architect can supply cost information for reflective or tinted glazing and film. A mechanical engineer can calculate the possible change in cooling load and its dollar value. For more information on film, see the Bibliography.

Rank

LOW MIDDLE: Examine costs and benefits carefully.

Summary and Conclusion:

Summary

The main points of this book can be summarized as follows:

1. Churches should act as good stewards of creation by using energy wisely. Energy conservation means saving energy and saving money.
2. Churches have unique opportunities for saving energy.
3. The best energy conservation measures vary from building to building, but as a rule, they fall into three main strategies:
   a. Don't send it up the chimney. Increase the efficiency of the heating and air conditioning system.
   b. Use it when and where you need it. Reduce the demand for heat by reducing the temperature difference between inside and outside and by controlling ventilation.
   c. Keep it where you want it. Increase the resistance to the flow of heat between inside and outside.
4. Strategies (a) and (b) usually save more and cost less than strategy (c). Therefore, investigate (a) and (b) first, even though insulation, storm windows, and other (c)-type measures generally come to mind first when energy conservation is mentioned.

Conclusion

This book is designed to help church officials make decisions about energy conservation and to enable them effectively to implement those decisions. Its main purpose is to guide churches by providing enough information for sound decisions. Decisions must be made by individual churches based on individual circumstances. However, this book will have missed its mark entirely if too much effort is misplaced in meetings debating whether or not caulking should be done before weather-stripping; that effort would be better spent doing the caulking or weather-stripping. As fuel prices go up and fuel becomes scarce, all these measures will be worthwhile. So start acting now, and keep saving energy.

The energy conservation measures discussed are not exclusive. Others could be added to the list, such as switching to a cheaper fuel or to a renewable fuel like wood or solar energy. Some churches should even consider moving to another building. Another form of energy conservation is sharing energy with others by using the building for more activities. For instance, if there is spare office space that must be heated anyway, can it be used by a local charity? Even if the sharing requires more heat, it is still better to heat a room to 65 degrees and use it than to heat it to 55 degrees and waste both heat and space.

Energy conservation does not stop at the church building. Once the church's conservation program has gathered momentum, look for opportunities to use these resources elsewhere. For instance, does the church own other buildings, such as a parsonage, where energy can be saved? Why not arrange for furnace tune-ups and other energy conservation measures to be implemented in homes of elderly or disabled church members? Do church representatives sit on boards of other organizations, such as the YMCA, a nursing home, or an elderly housing complex? The church can use its experience to get these organizations moving on energy conservation. Should the church sponsor an energy conservation course for homeowners or help develop car pools? In other words, energy conservation is a very real way of helping people and serving the community, as well as good stewardship of the church's own resources.

Appendix A explains life cycle costing. This is an appropriate method for comparing present and future costs in most church buildings. It is useful for comparing energy conservation measures where a decision must be made to implement some and not others and where economics is an important factor in the decision. Economics should never be the sole factor, nor is life cycle costing the only way to look at the economics. Many people in business prefer to look at the economic choices by at least three other methods: payback period, period to positive savings, and projected cash flows. Life cycle costing is presented here so that churches will have at least one valid method for economic comparison. It is particularly relevant for churches and other institutions that have a reasonably long-term view of the future, because it shows dramatically the long-term benefits of current investments in energy conservation. Although the explanation seems, at first glance, to be long and involved, the method is easily mastered and can be applied simply and quickly. A church may find it useful for other types of investment as well.

Most of the references in the Bibliography should be available from the local library. Use them to find out more about choosing materials or methods, or estimating energy savings, or simply to pursue interest. These are not the only sources, however; ask the librarian for more if necessary. Also, books are not the only resources. Check with architects, engineers, building contractors, and similar specialists. Salespeople can often provide good information, but get the other side of the story from competitors, customers, building inspectors, and the Better Business Bureau. A great deal of information can be obtained free or at low cost from federal and state governments. Many localities (and some denominations) have energy conservation paraprofessionals available for assistance.

Churches have a significant opportunity and responsibility to conserve energy in their own buildings and elsewhere. The real action is up to the church; only the church can reduce the energy it uses. Good stewardship starts at home: "First take the plank out of your own eye . . . ."

Appendixes:

Appendix A

Life Cycle Costing*

General Principles

Life Cycle Costing is a method of making economic evaluations. As the name implies, it is a way in which total costs over the course of the expected life of a system, device, or product can be estimated.

It is intuitively understandable that first costs are not the only determinant in making economic judgments between two or more options. It is clear that, in some cases, a higher first cost is justifiable if there are greater savings throughout the life of a product. For example, most people would agree that a more expensive refrigerator expected to last fifteen years is a "better buy" than a slightly less expensive one which only lasts five years.

Factors such as the expected cost of maintaining the product over its life and annual operating costs are commonly considered in everyday decision-making. Given two refrigerators, equal in other ways, one using less electricity every year is clearly "worth" some additional first cost. This is a simple example of the concept of life cycle costing; complexities come in attempting to make precise distinctions between the cost of various options over their expected lives. The method of life cycle costing described here is a way to refine these distinctions.

In making determinations of the relative cost-effectiveness of various weatherization, energy conservation, or solar energy applications, the factors to be considered are the following:

1. What is the first cost of the application?
2. How long will it last?
3. What is the cost of maintenance and repair throughout the life of the application?
4. What fuel cost savings can be expected from this application?
5. What are the operating costs?
6. What salvage value will there be at the end of the life cycle?

Present Value

In order to make comparisons between applications which have different lengths of useful life (and in order to make judgments based on a consistent set of values), all dollar costs must use the same value for each dollar. It is convenient to use the current value of the dollar as a constant. Since one can assume that inflation will continue, it is necessary to translate costs anticipated in the future to the present value of those future costs.

For example, a given machine is known to need replacement of a part after eight years of use, and, it is known that today, this replacement could cost $15.00; if it is assumed that the general rate of inflation is 10% per year, for instance, then one can find (through an equation or tables) that the cost in eight years will be $31.91, due to inflation.

The other factor which enters the calculations, is the general interest rate paid by banks. If it is known that $31.91 will be required in eight years, how much money should be put in a bank now in order to have the required amount in eight years? If the interest rate is assumed at 8%, for instance, then it will be found (from the same tables or equations) that $17.23 is the required investment. Therefore, one can say that the *present value* of that future cost is in fact $17.23, for that is the amount of resources which must be committed now in order to have the required funds in the future.

One can see that if the interest rate equalled the inflation rate (if the value of money put in a bank grew at the same rate that inflation degraded the value of that money), the two rates would cancel each other out and the present value of a future cost would be the same as today's cost for that item.

In the same way that the previous example was calculated, all costs for the proposed life (and salvage value, if included) of an energy conservation application can be calculated. Similarly, the benefits gained annually for energy saved as a result of the proposed application can be translated into the present value of the total energy savings through the life of the application. In this case, the inflation rate for fuel costs should be used, which may be different (higher) than the general inflation rate.

Once the present value of all of the costs and savings are found for the life of the application, the total future costs can be subtracted from the total future savings in energy costs. This yields the *net* present value of the savings.

Benefit/Cost Ratios

If one compares the "net present value of savings" with the "first cost" of the application, a benefit/cost ratio can be determined. For example, if from the previous calculations, a net present value of savings is found to be $67.50 for an application for which the first cost is $45.00, the benefit/cost ratio ($67.50 divided by $45.00) is 1.5; that is $1.50 of value for each $1.00 invested. This ratio is then comparable to that found for any other application, regardless of the length of expected life.

The major advantage of life cycle costing, compared to other economic evaluation methods, is that it accommodates the factor of inflation in the economy. Also differences in inflation rates between fuel costs and general inflation, and differences between inflation rates and interest rates can be accommodated in the calculations.

Assumptions

It should be noted that the calculations rest on a number of assumptions. Some of these—first cost estimates, maintenance requirements and major repair/replacement costs, salvage value, length of useful life and timing of maintenance and repair—are subject to normal errors in estimation. Other factors, in particular fuel inflation rates, general inflation rates, and interest rates, are assumptions which become, essentially, policy decisions reflecting an attitude concerning the future directions of the economy in general and fuel prices in particular. When beginning the use of life cycle costing as a tool, it may be desirable to experiment with the effects of various assumptions for these rates to become familiar with the implications of different "assumed futures."

First Costs

The total cost of materials [given in this article] for a particular application is the figure used as the "first cost" of that application. Figures for labor, profit, and overhead are not necessarily included in this "first cost" as they would be in a free market situation. Clearly, the benefit/cost ratio of an application will be down-graded the higher the first costs are. Therefore, benefit/cost ratios [in this article] will be comparatively high. In a normal market situation which includes labor, profit, and overhead costs, any proposed project with a benefit/cost ratio less than 1.00 would not be considered, as it would mean an economic loss.

In order to make the benefit/cost ratio more meaningful, it may be desirable to include realistic figures for labor, profit, and overhead in the life cycle costing of the applications, thus making the values comparable to nonsubsidized conditions. The result will be lower benefit/cost ratios. The threshold level, below which applications are considered infeasible, can be established as a policy decision and may be below 1.0

Formulae, Tables

Present Value. The present-day worth of a dollar to be received or spent one year from now is equal to that fraction of a dollar that, if invested today, would grow in value to one dollar in just one year. Assuming money can be invested at an interest rate (i), compounded annually, the present value of a dollar to be received or spent (n) years from now is calculated as:

Formula #1

$$PV_{\text{one dollar}} = \frac{1}{(1 + i)^n}$$

For example, the present value of $25.00 expenditure to be made five years from now, with an assumed interest rate of 8 percent, would be calculated, using the formula, as follows:

$$PV = 25.00 \times \frac{1}{(1 + 0.08)^5}$$

$$PV = 25.00 \times \frac{1}{1.469}$$

$$PV = 25.00 \times 0.6806$$

$$PV \doteq 17.02$$

The present value of $25.00 expenditure made five years from now, with an interest rate of 8%. is $17.02.

Figures of present value for different interest rates and numbers of years can also be found in the table on page 71.

Table # 1

n	4%	5%	6%	7%	8%	10%	12%	15%	20%
1	0.96154	0.95238	0.94340	0.9346	0.9259	0.9091	0.8929	0.8696	0.8333
2	0.92456	0.90703	0.89000	0.8734	0.8573	0.8264	0.7972	0.7561	0.6944
3	0.88900	0.86384	0.83962	0.8163	0.7938	0.7513	0.7118	0.6575	0.5787
4	0.85480	0.82270	0.79209	0.7629	0.7350	0.6830	0.6355	0.5718	0.4823
5	0.82193	0.78353	0.74726	0.7130	0.6806	0.6209	0.5674	0.4972	0.4019
6	0.79031	0.74622	0.70496	0.6663	0.6302	0.5645	0.5066	0.4323	0.3349
7	0.75992	0.71068	0.66506	0.6227	0.5835	0.5132	0.4523	0.3759	0.2791
8	0.73069	0.67684	0.62741	0.5820	0.5403	0.4665	0.4039	0.3269	0.2326
9	0.70259	0.64461	0.59190	0.5439	0.5002	0.4241	0.3606	0.2843	0.1938
10	0.67556	0.61391	0.55839	0.5083	0.4632	0.3855	0.3220	0.2472	0.1615
11	0.64958	0.58468	0.52679	0.4751	0.4289	0.3505	0.2875	0.2149	0.1346
12	0.62460	0.55684	0.49697	0.4440	0.3971	0.3186	0.2567	0.1869	0.1122
13	0.60057	0.53032	0.46884	0.4150	0.3677	0.2897	0.2292	0.1625	0.0935
14	0.57748	0.50507	0.44230	0.3878	0.3405	0.2633	0.2046	0.1413	0.0779
15	0.55526	0.48102	0.41727	0.3624	0.3152	0.2394	0.1827	0.1229	0.0649
16	0.53391	0.45811	0.39365	0.3387	0.2919	0.2176	0.1631	0.1069	0.0541
17	0.51337	0.43630	0.37136	0.3166	0.2703	0.1978	0.1456	0.0929	0.0451
18	0.49363	0.41552	0.35034	0.2959	0.2502	0.1799	0.1300	0.0808	0.0376
19	0.47464	0.39573	0.33051	0.2765	0.2317	0.1635	0.1161	0.0703	0.0313
20	0.45639	0.37689	0.31180	0.2584	0.2145	0.1486	0.1037	0.0611	0.0261
21	0.43883	0.35894	0.29416	0.2415	0.1987	0.1351	0.0926	0.0531	0.0217
22	0.42196	0.34185	0.27751	0.2257	0.1839	0.1228	0.0826	0.0462	0.0181
23	0.40573	0.32557	0.26180	0.2109	0.1703	0.1117	0.0738	0.0402	0.0151
24	0.39012	0.31007	0.24698	0.1971	0.1577	0.1015	0.0659	0.0349	0.0126
25	0.37512	0.29530	0.23300	0.1842	0.1460	0.0923	0.0588	0.0304	0.0105
26	0.36069	0.28124	0.21981	0.1722	0.1352	0.0839	0.0525	0.0264	0.0087
27	0.34682	0.26785	0.20737	0.1609	0.1252	0.0763	0.0469	0.0230	0.0073
28	0.33348	0.25509	0.19563	0.1504	0.1159	0.0693	0.0419	0.0200	0.0061
29	0.32065	0.24295	0.18456	0.1406	0.1073	0.0630	0.0374	0.0174	0.0051
30	0.30832	0.23138	0.17411	0.1314	0.0994	0.0573	0.0334	0.0151	0.0042

This table (or the equation) is used twice; to get an "interest rate constant" and an "inflation rate constant" both used in calculating present values of one-time repair and replacement costs which are scheduled to occur at given points in the life of the application. Similarly, present value of salvage at the end of the life cycle is calculated using this table.

To determine the present value of a series of annual events (of a constant nature) such as annual maintenance costs or annual fuel savings, totaled for the entire life cycle period, the following equation can be used:

Formula #2

$$P_e = F\frac{a(a^n - 1)}{a - 1} \text{ where } a = \frac{1 + g}{1 + i}$$

In this equation, the current annual expense or savings (F) is multiplied by a factor which takes into account the number of years in the life cycle (n), the inflation rate (g), and the interest rate (i).

For example, the present value of a $2.00 annual cost (for savings) over a five-year life, with interest rate assumed at 7% and inflation assumed at 8%, can be calculated with the formula as follows:

$$P_e = 2.00 \times \frac{a(a^5 - 1)}{a - 1} \qquad \text{where } a = \frac{1 + 0.08}{1 + 0.07}$$

Therefore, a = 1.0093, so,

$$P_e = 2.00 \times \frac{1.0093\ [(1.0093)^5 - 1]}{1.0093 - 1}$$

$$P_e = 2.00 \times \frac{1.0093\ (1.0474 - 1)}{0.0093}$$

$$P_e = 2.00 \times \frac{1.0093\ (0.0474)}{0.0093}$$

$$P_e = 2.00 \times \frac{0.0478}{0.0093}$$

$$P_e = 2.00 \times 5.1441$$

$$P_e = 10.29$$

Therefore, the present value for a five-year series of $2.00 annual costs (or savings), with interest rate of 7% and inflation at 8%, is $10.29.

Since this equation represents a relationship among three factors (life cycle period, interest rate, and inflation rate), it can be expressed as a series of tables, one table for each proposed life cycle period, which yield a multiplier given the interest and inflation rates. This multiplier is used with the current cost or savings to arrive at the present value of the total cost or savings over the life cycle period.

The tables below have been "rounded-off' to two decimal places. This results in a plus-or-minus 1% error inherent in the factors found from the chart. This is felt to be acceptable in that the repair and maintenance costs, the annual energy savings, the life cycle period, and the interest and inflation rates are all *estimated* figures used in the life cycle costing process described.

Table #2

2-Year Life Cycle

INTEREST RATE (i)	Inflation Rate (g)						
	6%	7%	8%	9%	10%	11%	12%
4%	2.06	2.09	2.12	2.15	2.18	2.21	2.24
5%	2.03	2.06	2.09	2.12	2.15	2.17	2.20
6%	2.00	2.03	2.06	2.09	2.11	2.14	2.17
7%	1.97	2.00	2.03	2.06	2.08	2.11	2.14
8%	1.94	1.97	2.00	2.03	2.06	2.08	2.11
9%	1.92	1.95	1.97	2.00	2.03	2.06	2.08

5-Year Life Cycle

INTEREST RATE (i)	Inflation Rate (g)						
	6%	7%	8%	9%	10%	11%	12%
4%	5.30	5.45	5.61	5.77	5.93	6.10	6.28
5%	5.14	5.29	5.45	5.60	5.76	5.93	6.09
6%	5.00	5.14	5.29	5.44	5.60	5.75	5.92
7%	4.86	5.00	5.14	5.29	5.44	5.59	5.75
8%	4.73	4.86	5.00	5.14	5.28	5.43	5.58
9%	4.60	4.73	4.86	5.00	5.14	5.28	5.43

7-Year Life Cycle

INTEREST RATE (i)	Inflation Rate (g) 6%	7%	8%	9%	10%	11%	12%
4%	7.56	7.86	8.16	8.48	8.82	9.16	9.52
5%	7.27	7.55	7.85	8.15	8.47	8.80	9.14
6%	7.00	7.27	7.55	7.84	8.14	8.45	8.78
7%	6.74	7.00	7.27	7.54	7.83	8.13	8.44
8%	6.50	6.75	7.00	7.26	7.54	7.82	8.12
9%	6.27	6.50	6.75	7.00	7.25	7.53	7.81

10-Year Life Cycle

INTEREST RATE (i)	Inflation Rate (g) 6%	7%	8%	9%	10%	11%	12%
4%	11.12	11.73	12.38	13.06	13.79	14.56	15.37
5%	10.54	11.11	11.71	12.35	13.03	13.75	14.51
6%	10.00	10.53	11.10	11.70	12.33	13.00	13.71
7%	9.50	10.00	10.53	11.09	11.68	12.30	12.97
8%	9.04	9.50	10.00	10.52	11.08	11.66	12.28
9%	8.60	9.04	9.51	10.00	10.52	11.07	11.65

15-Year Life Cycle

INTEREST RATE (i)	Inflation Rate (g) 6%	7%	8%	9%	10%	11%	12%
4%	17.53	18.97	20.56	22.29	24.19	26.27	28.55
5%	16.20	17.50	18.93	20.49	22.21	24.08	26.13
6%	15.00	16.18	17.48	18.89	20.43	22.12	23.97
7%	13.93	15.00	16.17	17.45	18.85	20.37	22.04
8%	12.96	13.94	15.00	16.16	17.43	18.81	20.31
9%	12.09	12.98	13.94	15.00	16.15	17.40	18.77

20-Year Life Cycle

INTEREST RATE (i)	Inflation Rate (g) 6%	7%	8%	9%	10%	11%	12%
4%	24.58	27.32	30.43	33.96	37.96	42.49	47.63
5%	22.13	24.53	27.24	30.31	33.78	37.71	42.17
6%	20.00	22.10	24.48	27.16	30.19	33.61	37.48
7%	18.15	20.00	22.08	24.43	27.08	30.07	33.44
8%	16.53	18.16	20.00	22.06	24.39	27.00	29.95
9%	15.11	16.56	18.18	20.00	22.04	24.34	26.92

35-Year Life Cycle

INTEREST RATE (i)	Inflation Rate (g) 6%	7%	8%	9%	10%	11%	12%
4%	50.23	60.83	74.16	90.98	112.23	139.15	173.32
5%	41.70	50.05	60.50	73.60	90.08	110.88	137.15
6%	35.00	41.63	49.88	60.17	73.05	89.22	109.56
7%	29.63	35.00	41.56	49.71	59.85	72.51	88.38
8%	25.45	29.73	35.00	41.50	49.54	59.53	71.99
9%	22.03	25.52	29.78	35.00	41.43	49.37	59.23

Evaluation Process

To use life cycle costing to compare the value of the applications, the following procedure may be used. The result is a benefit/cost ratio for each application.

I. Inputs

1. General assumptions—
   a. Inflation rate of fuel
   b. General inflation rate
   c. Interest rate
2. Information specific to the application—
   a. Life cycle period
   b. Estimated first cost
   c. Estimated annual maintenance and operating cost

d. Schedule of repairs estimated and costs
   (Note: These are expressed in terms of an expense at a certain number of years into the life cycle, for example, $15.00 must be spent eight years from the implementation of the application. Note also that repeating maintenance costs must be calculated for each time they occur. For example, a maintenance item which is scheduled to occur every five years on an application with a twenty-year life cycle, must be calculated three times, for year five, year ten, and year fifteen of the application's life cycle.)
e. Estimated fuel savings for the first year of the application's use
f. Salvage value

II. Operations

1. *Determine present value of future repair and replacement costs.*
   a. Using equation 1 or Table 1, find both the interest constant and the general inflation constant for the appropriate year of the life cycle for the particular repair or replacement event, and multiply as follows:

$$\text{Cost for repair} \times \frac{\text{Interest Constant}}{\text{Inflation Constant}} = \begin{matrix}\text{Present}\\ \text{Value}\end{matrix}$$

This should be repeated for each repair or replacement item on the schedule, with constants found for each of the necessary years into the life cycle.

Example: If schedule calls for a $15.00 repair cost at year eight of the life cycle, and the interest rate is assumed at 8%, a general inflation is assumed at 10%, then (using Table 1 for constants)

$$15 \times \frac{.54}{.47} = 17.23 \text{ (dollars present value)}$$

   b. Total the present values of all repair and replacement costs.

2. *Determine the present value of the salvage value.* Using equation 1 or Table 1, find both the interest constant and the general inflation constant for the last year in the life cycle and multiply by the estimate of the salvage value if the parts were salvaged today.

$$\text{Salvage value} \times \frac{\text{Interest Constant}}{\text{Inflation Constant}} = \begin{matrix}\text{Present}\\ \text{Value}\end{matrix}$$

Example: If it is determined that, if the device had completed its life today, it would have a salvage value of $7.50, and the life of the device is estimated at ten years, and the same interest and general inflation rates as the previous example are assumed, then (using Table 1 for constants)

$$7.50 \times \frac{.46}{.38} = 9.08 \text{ (dollars present value)}$$

3. *Determine the present value for regular, annual, maintenance, and operating costs.* From equation 2, or from the appropriate Table 2 for the application's life cycle period, find the appropriate constant, given the general inflation and interest rates assumed, and multiply by the annual cost, as follows:

$$\text{Annual Maint. \& Oper. Cost} \times \text{Constant} = \begin{matrix}\text{Present}\\ \text{Value}\end{matrix}$$

   Example: If annual maintenance of a device expected to last 10 years is estimated at $2.00, the general inflation rate is assumed to be 6%, and the interest rate is assumed to be 8%, then (using Table 2 for 10-year life cycle):

   $2.00 \times 9.04 = 18.08$ (dollars Present Value)

4. *Determine present value of future fuel cost savings.* From equation 2, or from the appropriate Table 2 for the application's life cycle period, find the appropriate constant given the *fuel* cost inflation rate and the interest rate assumed. Multiply by the annual fuel cost savings found for the application as follows:

$$\text{Annual fuel cost savings} \times \text{constant} = \begin{matrix}\text{Present}\\ \text{Value}\end{matrix}$$

   Example: Given the same interest rate assumption as in the previous example, a fuel cost inflation rate of 10%, and an annual fuel savings (found from energy savings section for a particular application) of $8.62, then:

   $8.62 \times 11.08 = 95.51$ (dollars present value)

5. *Determine Benefit/Cost Ratio (Summary)*
   a. Add the present values of repair and replacement costs to the present value of annual maintenance and operating cost. Subtract this total from the present value of fuel cost savings.

$(PV_{salvage} + PV_{fuel\ savings}) - (PV_{repair} + PV_{maintenance} + PV_{operation}) =$ Net Present Value of Savings

Example: Using numbers found in previous examples.
(9.08 + 95.51) − (17.23 + 18.08) = $69.28 (Net PV)

b. Divide net present value by estimated first cost,

$$\frac{\text{Net PV}}{\text{First Cost}} = \text{Benefit/Cost Ratio}$$

This is the value of benefit gained for each dollar spent.

Example: Using previous example and estimated first cost of $40.00,

$$\frac{69.28}{40.00} = 1.73 \text{ (Benefit/Cost)}$$

The benefit/cost ratios of particular applications for a particular church can be used in deciding which of several possible applications would be the most cost-effective for that church.

Example 1

Finally, complete examples are given below for two imagined applications:

Inputs

1. General assumptions:
   a. Inflation rate of fuel 10%
   b. General inflation rate 8%
   c. Interest rate 6%
2. Specific information on application—4 insulating shutters
   a. Life cycle period 20 yrs.
   b. Estimated first cost $75.00
   c. Est. annual maintenance 0
   d. Schedule of repairs & costs
      At 5 years, repairs costing $11.20 Total

e. Est. fuel cost savings for 1st year — $4.16
f. Salvage value — 0

Operations

1. Determine present value of future repair and replacement costs.
   a. (Using equation 1 or Table 1 twice for each calculation)

      $\$11.20 \times \frac{.75}{.68} =$ $12.35

   b. (total) = $12.35
2. Determine present value of salvage value. None
3. Determine present value for annual maintenance and operating costs. None
4. Determine the present value of future fuel cost savings (using equation 2 or Table 2 for 20-year life cycle)

   $$\text{Constant} = \frac{a\,(a^{20} - 1)}{a - 1} \text{ where } a = \frac{1 + .10}{1 + .06}$$

   $\$4.16 \times 30.19 = \$125.59$

5. Determine benefit/cost ratio
   a. Find net present value of savings

(future fuel cost savings)	−	(total future costs)	=	net PV savings
125.59	−	12.35	=	113.24

   b. Net PV divided by first cost = benefit/cost

   $$\frac{113.24}{75.00} = 1.51$$

Example 2

Inputs

1. General assumption
   a. Inflation rate of fuel — 10%
   b. General inflation rate — 8%
   c. Interest rate — 6%
2. Specific information on application—Lean-to Air/Rock Collector
   a. Life cycle period — 20 years
   b. Estimated first cost — $750.00

c. Est. annual maintenance $4.00/yr.

d. Schedule of repairs and costs

  at 5 yrs., a repair costing $ 5.00

  at 10 yrs., repairs costing $20.00

  at 15 yrs., a repair costing $15.00

e. Est. fuel cost savings for 1st year $75.00

f. Salvage value 0

Operations

1. Determine present value of future repair and replacement costs.
   a. (Using equation 1 or Table 1 twice for each calculation)

$$5.00 \times \frac{.75}{.68} = \$\ 5.51$$

$$20.00 \times \frac{.56}{.46} = \$\ 24.35$$

$$15.00 \times \frac{.42}{.32} = \$\ 19.69$$

   b. (total) = $ 49.55

2. Determine present value of salvage value. None
3. Determine present value for annual maintenance and operating costs (using equation 2 or Table 2 for 20-year life cycle)

$$\text{Constant} = \frac{a\,(a^{20} - 1)}{a - 1} \text{ where } a = \frac{1 + .08}{1 + .06}$$

$$4.00 \times 24.48 = \$\ 97.92$$

4. Determine the present value of future fuel cost savings (using equation 2 for 20-year life cycle)

$$\text{Constant} = \frac{a\,(a^{20} - 1)}{a - 1} \text{ where } a = \frac{1 + .10}{1 + .06}$$

$$\$75.00 \times 30.19 = \$2{,}264.25$$

5. Determine Benefit/Cost Ratio
   a. Find net present value of savings

(PV of future fuel cost savings minus PV of future costs)

$2,264,25 – (49.55 + 97.92) = $2,116.78

   b. (Net PV divided by first cost) = benefit/cost

$$\frac{2{,}116.78}{750.00} = \$\ 2.82$$

Appendix B MONITORING ENERGY USE

YEAR 19 ___	AMOUNT OF FUEL *	COST * PER UNIT	TOTAL $
JANUARY			
FEBRUARY			
MARCH			
APRIL			
MAY			
JUNE			
JULY			
AUGUST			
SEPTEMBER			
OCTOBER			
NOVEMBER			
DECEMBER			
TOTAL			

*Fuel Oil—Amount: Gallons; Cost:$/Gal.
Electricity—Amount: Kilowatt Hours; Cost: $/KWH
Natural Gas—Amount: 100s of Cubic Feet; Cost: $/Cu. Ft.

Bibliography:

Bibliography

Energy Conservation Philosophy, Theology, Theory

Hayes, Denis, "Energy: The Case for Conservation." Worldwatch Paper 4, January 1976, Worldwatch Institute, 1776 Massachusetts Avenue, N. W., Washington, D. C. 20036.

Lovins, Amory, "Energy Strategy: The Road Not Taken?" Friends of the Earth's *Not Man Apart,* November 1976.

"Energy Ethics Consultation." Report of the National Council of Churches, 475 Riverside Drive, Room 572, New York, New York 10027, October 1977.

Heating and Cooling

ASHRAE Handbook and Product Directory, 1977 Fundamentals. American Society of Heating, Refrigerating, and Air-Conditioning Engineers, Inc., 1977.

Leckie, Jim et al, eds., *Other Homes and Garbage.* San Francisco, Calif.: Sierra Club Books, 1975.

McGuinness, William and Benjamin Stein, *Mechanical and Electrical Equipment for Buildings,* Fifth Edition. New York: John Wiley and Sons, 1971.

Energy Conservation

General:

Dubin, F. S. et al. *How to Save Energy and Cut Costs in Existing Industrial and Commercial Buildings: An Energy Conservation Manual.* Park Ridge, N. J.: Noyes Data Corp., 1976.

Total Environmental Action, Inc. *Energy Stewardship: Energy Conservation Analysis of Three Massachusetts Churches.* Harrisville, N. H.: TEA, Inc., 1977.

U.S. Department of Housing and Urban Development. *In the Bank . . . or Up the Chimney?,* April 1975.

Insulation:

Federal Energy Administration. *Project Retro-Tech,* Teacher's Kit for Course on Home Weatherization, May 1976.

Homes/Apartments, *Insulation Manual.* Rockville, Md.: NAHB Research Foundation, September 1971.

"The Overselling of Insulation," *Consumer Reports,* November 1977.

Caulking and Weather Stripping:

Day, "Common Sense Guide to Caulks and Caulking," *Popular Science,* September 1974.

"Exterior Caulking Compounds," *Consumer Reports,* 1977 (Buying Guide).

"Weatherstripping," *Consumer Reports,* February 1977.

Storm Windows:

"Consumer's Guide to Storm Windows," *Mechanics Illustrated,* Vol. 69, October 1973, pp. 98-99.

Sickler, "Low-Cost Storm Windows for Basement Wells," *Popular Science,* Vol. 205, October 1974, p. 164.

"Storm Windows and Doors," *Changing Times,* Vol. 28, October 1974, pp. 21-23.

"This Storm Window Installs on the Inside," *Popular Mechanics,* Vol. 143, February 1975, p. 140.

Walton, "All-Plastic Storm Windows You Assemble in Minutes," *Popular Mechanics,* Vol. 142, December 1974, p. 36.

Weinsteiger, "Five Dollar Storm Window," *Organic Gardening and Farming,* Vol. 24, October 1977, p. 150.

Reflective Glass and Films:

Powell, "Solar Ban Film and Glass Keeps Your Home or Recreational Van Cool," *Popular Science,* Vol. 210, May 1977, p. 144.

Smith, "Energy-Saving Window: Reflective Glass," *Popular Science,* Vol. 209, November 1976, p. 176.

Solar Energy

Anderson, Bruce N., *Solar Energy: Fundamentals in Building Design.* New York: McGraw-Hill, 1977.

———, *The Solar Home Book: Heating, Cooling, and Designing with the Sun.* Harrisville, N. H.: Cheshire Books, 1976.

Scully, Prowler, and Anderson, *The Fuel Savers: A Kit of Solar Ideas for Existing Homes.* Harrisville, N. H.: Total Environmental Action, Inc., 1975.